Pottery Production, Settlement Patterns and Development of Social Complexity in the Yuanqu Basin, North-Central China

Xiangming Dai

BAR International Series 1502
2006

Published in 2016 by
BAR Publishing, Oxford

BAR International Series 1502

Pottery Production, Settlement Patterns and Development of Social Complexity in the Yuanqu Basin, North-Central China

ISBN 978 1 84171 939 9

BAR Publishing is the trading name of British Archaeological Reports (Oxford) Ltd. British Archaeological Reports was first incorporated in 1974 to publish the BAR Series, International and British. In 1992 Hadrian Books Ltd became part of the BAR group. This volume was originally published by Archaeopress in conjunction with British Archaeological Reports (Oxford) Ltd / Hadrian Books Ltd, the Series principal publisher, in 2006. This present volume is published by BAR Publishing, 2016.

Printed in England

BAR
PUBLISHING

BAR titles are available from:

BAR Publishing
122 Banbury Rd, Oxford, OX2 7BP, UK
EMAIL info@barpublishing.com
PHONE +44 (0)1865 310431
FAX +44 (0)1865 316916
www.barpublishing.com

Contents

List of Figures iii
List of Tables iv
Abstract vi
Acknowledgements vii

Chapter One: Introduction 1

1.1. Research Objective and Background 1
1.2. Location and Natural Environment of the Yuanqu Basin 3
1.3. Archaeological Setting and Chronological Framework for the Yuanqu Basin 3
1.4. Surveys, Excavations, and Studies for the Yuanqu Basin 5
1.5. Research Approach, Procedure, and Data 6
1.6. Summary 6

Chapter Two: Social Evolution and Settlement Patterns: A Discussion of Theories and Methods 8

2.1. Introduction 8
2.2. Uni-linear vs. Multi-linear Evolution: Processes of Social Evolution 8
2.3. Mono-causal vs. Multi-variate Explanations: Dynamics of Social Evolution 9
2.4. Settlement Archaeology: An Effective Approach to Social Complexity 10
2.5. Summary 12

Chapter Three: Settlement Patterns and Social Complexity in the Yuanqu Basin .13

3.1. Introduction 13
3.1.1. Approaches to Settlement Patterns 13
3.1.2. Procurement and Utilisation of Data 13
3.2. Peiligang Period 14
3.3. Early Yangshao Period 16
3.4. Middle Yangshao Period 19
3.5. Late Yangshao Period 23
3.6. Miaodigou II Period 25
3.7. Longshan Period 28
3.8. Erlitou Period 31
3.9. Erligang Period 33
3.10. Summary and Discussion 38

Chapter Four: Modelling Pottery Production: Theories, Methods, and Data 40

4.1. Introduction 40
4.2. Mode of production, Specialisation, and Social Complexity 40
  4.2.1. The Mode of Production and Socio-political Structure 40
  4.2.2. Specialisation 41
  4.2.3. A model for Assessing Specialisation or the Organisation of Production 42
4.3. Approaches to Identifying the Mode of Production or Specialisation 44
  4.3.1. Direct Evidence 44
  4.3.2. Indirect Evidence 44
  4.3.3. Standardisation, Diversity, and Specialisation 44
  4.3.4. Measurements of Standardisation, Diversity and Variation 45
4.4. Data Collection 46
  4.4.1. Sources of Ceramic Data 46
  4.4.2. Establishment of Analytical Units 47
  4.4.3. Classification of Ceramic Vessels 47
  4.4.4. Measurement and Recording of Pottery 48
  4.4.5. Sample Size 49
  4.5. Summary 50

Chapter Five: Pottery Production in a Tribal-level Society: A Case Study of the Early Yangshao Period at the Dongguan Site 52

5.1. Introduction 52
5.2. Direct Evidence of Pottery Production 52
5.3. Pottery Classification and the Assemblage Composition 55
5.4. Indirect Evidence of Pottery Production 57
5.4.1. Analyses of Nominal Data 57
5.4.2. Analyses of Metric Data 58
5.5. Summary 62

Chapter Six: Pottery Production in a Chiefdom-level Society: A Case Study of the Miaodigou II Period at the Dongguan Site 63

6.1. Introduction 63
6.2. Direct Evidence of Pottery Production 63
6.3. Pottery Classification and the Assemblage Composition 70
6.4. Indirect Evidence of Pottery Production 70
6.4.1. Analyses of Nominal Data 70
6.4.2. Analyses of Metric Data 77
6.5. Summary 81

Chapter Seven: Pottery Production in a State-level Society: A Case Study of the Erlitou &Erligang Period at the Nanguan Site 83

7.1. Introduction 83
7.2. Direct Evidence of Pottery Production 83
7.3. Pottery Classification and the Assemblage Composition 90
7.4. Indirect Evidence of Pottery Production 97
7.4.1. Analyses of Nominal Data 97
7.4.2. Analyses of Metric Data 99
7.5. Summary 105

Chapter Eight: Conclusions 106

8.1. Introduction 106
8.2. Trajectories to Social Complexity 106
8.3. Changes of Modes of Pottery Production and Development of Specialisation 106
8.4. Productive Specialisation and Social Complexity 108
8.5. Questions for the Future Research 109
8.6. Significance of This Study 110

Appendix 1. Neolithic and early Bronze Age sites in the Yuanqu Basin. 111
Appendix 2. The number and proportion of ceramic classes in the Dongguan I assemblage 113
Appendix 3. The number and proportion of ceramic classes in the Miaodigou II assemblage. 114
Appendix 4. The compositions of ceramic paste for the five ceramic classes in the Miaodigou II assemblage. 116
Appendix 5. The Compositions of ceramic decorations for the eight ceramic classes in the Miaodigou II assemblage. 117
Appendix 6. The number and proportion of ceramic classes in the Late Erlitou and Erligang assemblages. 118
Appendix 7. The compositions of ceramic paste for the five ceramic classes in the Late Erlitou and Erligang assemblages 119
Appendix 8. The number and proportion of pottery vessels by slow-wheel shaping for the five ceramic classes in the Late Erlitou and Erligang assemblages. 120

Bibliography 121

List of Figures

Figure 1-1. Location of the Yuanqu Basin in Shanxi and Henan. 2
Figure 1-2. Physiographic map of the Yuanqu Basin. 2
Figure 1-3. Regional divisions of the Neolithic cultures in China. 4
Figure 3-1. Distribution of archaeological sites in the Yuanqu Basin. 14
Figure 3-2. Distribution of Peiligang sites in the Yuanqu Basin. 15
Figure 3-3. Distribution of Early Yangshao sites in the Yuanqu Basin. 16
Figure 3-4. Histogram showing the frequency of site sizes of Dongguan I sites. 17
Figure 3-5. Rank-size distribution of Dongguan I sites. 17
Figure 3-6. Map of the Dongguan site, showing the distribution of excavation areas. 18
Figure 3-7. Distribution of some Dongguan I features in Area IV of Dongguan. 18
Figure 3-8. Distribution of Middle Yangshao sites in the Yuanqu Basin. 20
Figure 3-9. Histogram showing three levels of settlement hierarchy in the Yuanqu Basin, Middle Yangshao period. 20
Figure 3-10. Rank-size distribution of Middle Yangshao sites in the Yuanqu Basin. 21
Figure 3-11. An eagle-shaped tripod found at the Quanhu site in Huaxian, Shaanxi. 22
Figure 3-12. A large building (F105) discovered at Xipo in Lingbao, Henan. 22
Figure 3-13. Distribution of Late Yangshao sites in the Yuanqu Basin. 24
Figure 3-14. Histogram showing two levels of settlement hierarchy in the Yuanqu Basin, Late Yangshao period. 25
Figure 3-15. Rank-size distribution of Late Yangshao sites in the Yuanqu Basin. 25
Figure 3-16. Distribution of Miaodigou II sites in the Yuanqu Basin. 26
Figure 3-17. Histogram showing two levels of settlement hierarchy in the Yuanqu Basin, Miaodigou II period. 27
Figure 3-18. Rank-size distribution of Miaodigou II sites in the Yuanqu Basin. 27
Figure 3-19. Plan and cross-section of the pit IIIH63 at Dongguan, late Miaodigou II period. 28
Figure 3-20. Distribution of Longshan sites in the Yuanqu Basin. 29
Figure 3-21. Histogram showing three levels of settlement hierarchy in the Yuanqu Basin, Longshan period. 30
Figure 3-22. Rank-size distribution of Longshan sites in the Yuanqu Basin. 30
Figure 3-23. Plan of the tomb IM5 at Dongguan, Longshan period. 31
Figure 3-24. Distribution of Erlitou sites in the Yuanqu Basin. 32
Figure 3-25. Histogram showing two levels of settlement hierarchy in the Yuanqu Basin, Erlitou period. 32
Figure 3-26. Rank-size distribution of Erlitou sites in the Yuanqu Basin. 33
Figure 3-27. Outline of the Late Erlitou settlement at Nanguan. 33
Figure 3-28. Distribution of Erligang sites in the Yuanqu Basin. 34
Figure 3-29. Histogram showing two levels of settlement hierarchy in the Yuanqu Basin, Erligang period. 35
Figure 3-30. Rank-size distribution of Erligang sites in the Yuanqu Basin. 35
Figure 3-31. Map of the Erligang settlement at Nanguan. 36
Figure 3-32. Plan of the burial M16 at Nanguan, Lower Erligang period. 37
Figure 3-33. Plan and cross-section of the pit H353 at Nanguan, Lower Erligang Period. 37
Figure 4-1. Metric variables measured on Yuanqu vessels. 49
Figure 4-2. Distribution of CV values with different sample sizes for the orifice diameter of *li* tripods in the Lower Erligang period at Nanguan. 50
Figure 5-1. Anvil of the Dongguan I period discovered in the Dongguan site. 52
Figure 5-2. Distribution of Dongguan I kilns at the Dongguan site. 53
Figure 5-3. Distribution of Early Yangshao kilns at the Beishouling site. 54
Figure 5-4. Map of the Early Yangshao settlement at Jiangzhai, showing locations of kilns. 54
Figure 5-5a. Dongguan I ceramics from the Dongguan site. 55
Figure 5-5b. Dongguan I ceramics from the Dongguan site. 56
Figure 5-6. Distribution of ceramic classes in the Dongguan I assemblage. 57
Figure 6-1. Pottery-manufacturing tools of the Miaodigou II period from the Fengcun and Dongguan site. 63
Figure 6-2. Location and surface of the kilns Y501 and Y502 at Ningjiapo 64
Figure 6-3. Plan and cross-sections of the kiln Y501 at Ningjiapo. 64
Figure 6-4. Distribution of Miaodigou II kilns and relevant features at Area III of Dongguan. 66
Figure 6-5. Plan and cross-sections of the kiln IIIY4 at Dongguan. 67
Figure 6-6. Plan and cross-sections of the pit IIIH22 at Dongguan. 68

Figure 6-7. Plan and cross-sections of the pit IIIH11 at Dongguan. 69
Figure 6-8a. Early Miaodigou II ceramics from the Dongguan site. 71
Figure 6-8b. Early Miaodigou II ceramics from the Dongguan site. 72
Figure 6-9a. Middle Miaodigou II ceramics from the Dongguan site. 73
Figure 6-9b. Middle Miaodigou II ceramics from the Dongguan site. 73
Figure 6-10a. Late Miaodigou II ceramics from the Dongguan site. 74
Figure 6-10b. Late Miaodigou II ceramics from the Dongguan site. 74
Figure 6-11a. Distribution of ceramic classes in the early Miaodigou II assemblage 75
Figure 6-11b. Distribution of ceramic classes in the middle Miaodigou II assemblage. 75
Figure 6-11c. Distribution of ceramic classes in the late Miaodigou II assemblage. 75
Figure 7-1. Pottery-manufacturing tools of the Late Erlitou and Erligang periods from the Nanguan site. 83
Figure 7-2. Plan and cross-sections of the kiln Y6 at Nanguan. 84
Figure 7-3. Plan and cross-section of the kiln Y9 at Nanguan. 85
Figure 7-4. Map of the Nanguan site, showing the outline of the rammed-earth enclosure, positions of the palatial district and the Upper Erligang kilns. 86
Figure 7-5. Jade and bronze objects of the Erligang period at Nanguan. 88
Figure 7-6. Ritual ceramic wares from the tomb M40 of Upper Erligang at Nanguan. 89
Figure 7-7. Valuable pottery vessels of the Erligang period at Nanguan. 89
Figure 7-8a. Late Erlitou ceramics from the Nanguan site. 91
Figure 7-8b. Late Erlitou ceramics from the Nanguan site. 92
Figure 7-9a. Lower Erligang ceramics from the Nanguan site. 93
Figure 7-9b. Lower Erligang ceramics from the Nanguan site. 94
Figure 7-10a. Upper Erligang ceramics from the Nanguan site. 95
Figure 7-10b. Upper Erligang ceramics from the Nanguan site. 96
Figure 7-11a. Distribution of ceramic classes in the Late Erlitou assemblage. 96
Figure 7-11b. Distribution of ceramic classes in the Lower Erligang assemblage. 97
Figure 7-11c. Distribution of ceramic classes in the Upper Erligang assemblage. 97
Figure 7-12. Box plots showing variable ranges of CV values for Different chronological phases in Dongguan I, Miaodigou II and Erlitou & Erligang assemblages. 104
Figure 7-13. Box plots showing variable ranges of CV values for pooled data in Dongguan I, Miaodigou II and Erlitou & Erligang assemblages. 104

List of Tables

Table 1-1. The Cultural Sequence and Chronological Periodisation for the Central Plain Region. 4
Table 1-2. The Chronological Sequence of Archaeological Cultures for the Yuanqu Basin. 5
Table 4-1. Chronological phases and time span for the four periods. 47
Table 5-1. Sample means, standard deviations (SD), and coefficients of variation (CV) for the five classes of ceramics in the Dongguan I assemblage. 60
Table 5-2. Analysis of variance (ANOVA) for the five classes of ceramics Between the early and late phase in the Dongguan I assemblage. 60
Table 6-1. Frequency of pottery colours for the three phases of the Miaodigou II period at the Dongguan site. 76
Table 6-2. Frequency of primary paste categories for the three phases of the Miaodigou II period at the Dongguan site. 76
Table 6-3. Frequency of surface treatments and decorations for the three phases of the Miaodigou II period at the Dongguan site. 76
Table 6-4. Correlation coefficients among metric variables for the three ceramic classes in the Miaodigou II assemblage of the Dongguan site. 77
Table 6-5. Analysis of variance (ANOVA) for the four classes of ceramics between the three phases in the Miaodigou II assemblage of the Dongguan site. 78
Table 6-6. Sample means (M), standard deviations (SD), and coefficients of variation (C.V.) for the eight classes of ceramics in the Miaodigou II assemblage of the Dongguan site. 79
Table 7-1. Frequency of pottery colours for the three phases of the Erlitou and Erligang periods at the Nanguan site. 98
Table 7-2. Frequency of primary paste categories for the three phases of the Erlitou and Erligang periods at the Nanguan site. 98
Table 7-3. Frequency of surface treatments and decorations for the three phases of the Late Erlitou and Erligang periods at the Nanguan site. 98
Table 7-4. Correlation coefficients among metric variables for the five ceramic classes in the Erlitou and

Erligang assemblages of the Nanguan site. ... 99
Table 7-5a. Analysis of variance (ANOVA) for the four ceramic classes between the Late Erlitou and the Lower Erligang assemblages of the Nanguan site. ... 100
Table 7-5b. Analysis of variance (ANOVA) for the five ceramic classes between The Lower and Upper Erligang assemblages of the Nanguan site... 100
Table 7-6. Sample means, standard deviations (SD), and coefficients of Variation (C.V.) for the five classes of ceramics in the Late Erlitou and Erligang assemblages of the Nanguan site. ... 101

Abstract

This book investigates the development of social complexity and the changes of modes of pottery production from the Neolithic to early Bronze Age in the Yuanqu Basin, north-central China. The research focuses on the development of specialised pottery production in different societies.

Through settlement pattern studies, I examine the social changes during eight cultural periods from the Neolithic to the early Bronze Age (ca. 6200-1300 B.C.). The settlement analyses address 46 sites, including seven excavated sites in the Yuanqu Basin. The results suggest that the Yuanqu Basin may have witnessed a developmental process from tribal-level (local group), through chiefdom-level to state-level societies, with some fluctuations in population size, social scale, and degree of social integration. On the basis of these socio-political backgrounds, the discussion of pottery production concentrates on four periods: Early Yangshao (ca. 4600-4100), Miaodigou II (ca. 2900-2500), and Late Erlitou (ca. 1800/1700-1600/1500 B.C.) and Erligang (ca. 1600-1300 B.C.), which can be identified as tribe-, chiefdom-, and state-level societies. I assess pottery production from four parameters: concentration, scale, intensity, and context of production, through direct and indirect evidence, mainly based on the data from two excavated sites (Dongguan and Nanguan). The analytical results suggest that the mode of pottery production in the Basin may have experienced a change from a dispersed, small-scale and part-time household industry and/or an individual workshop industry to concentrated, large-scale and full-time nucleated workshops. The social context of production possibly experienced a change from kin-group organised specialisation, independent specialisation to elite-involved specialisation. Correspondingly, the promotion of specialisation of pottery production is demonstrated by increasing standardisation in nominal and metric variables of pottery.

This study argues that the initial and low-level specialisation might occur in the simple and non-stratified society, and the dramatic change in the mode of pottery production or the degree of specialisation is not necessarily related to the emergence and development of chiefdom-level societies, but rather associated with the formation of state-level societies, as demonstrated by the Yuanqu case.

Acknowledgements

At the end of 1999, I attended an archaeological lecture at the Institute of Archaeology, Chinese Academy of Social Science (IACASS), which was given by Professor Tim Murray, Head of the School of Historical and European studies, La Trobe University. Afterwards, I asked several questions about the archaeological theories associated with his lecture. Several months later, in June of 2000, Professor Chen Xingcan from IACASS forwarded a message to me: Professor Murray would like to provide me a chance to undertake PhD study at La Trobe University. This information was completely unexpected for me. I knew that this would not only be a new opportunity, but also a big challenge for my academic career. Professor Chen enthusiastically encouraged me to grasp this opportunity. After an unusual process, in July of 2002, I became a PhD student at the Archaeology Department, La Trobe University, and then spent an enjoyable time in Melbourne, Australia. I would like to express my sincere appreciation to Professor Murray. Without his generous help and support, it would have been impossible for me to obtain this precious opportunity to study at La Trobe University.

I owe a deep debt of gratitude to Dr. Li Liu and Dr. David Frankel, my dissertation supervisors.

Over the past three years, Dr. Liu has given me tremendous help and support. She has always encouraged and helped me to overcome various difficulties in both study and life. I have learned much in both Western and Chinese archaeology from her. Her comments and advice on my thesis writing have made me avoid a lot of inappropriate expressions, and always led me to a deeper understanding of archaeological studies.

Dr. Frankel has spent a large amount of time in reviewing my thesis. He has not only given many valuable suggestions on settlement and pottery analyses, but also corrected my errors in English with great patience.

I am thankful to Dr. Phillip Edwards and Dr. Gyoung-Ah Lee, who read several chapters of this thesis, and provided a great deal of advice about this study.

I am deeply indebted to Mr. Wei Ming, who has always been very helpful whenever I have had any questions in using computer programs. I have also got much valuable advice and help in life from him throughout the past three years.

I would like to use this opportunity to express my gratitude to Ms. Stella Bromilow, Ms. Rose Allen, Mr. Rudy Frank and all other staff members of the Archaeology Department, for their various help during my study at La Trobe University.

I am very grateful to the late Professor Yu Weichao of the National Museum of China, Professor Yan Wenming of Beijing University, and Dr. Yun Kuen Lee of Harvard University, who provided the references for my application for doctoral candidature. I have also received enthusiastic encouragement and intellectual help from Professor Yu and Professor Yan for many years in my academic career. I have deeply missed Professor Yu who unfortunately died in late 2003.

I am very thankful to Professor Zhu Fenghan, Professor Li Ji, Professor Wang Xiaotian, and Professor Dong Qi, directors of the National Museum of China (the former National Museum of Chinese History), who have supported my study at La Trobe University and preserved my position in the Museum. I am especially indebted to Professor Xin Lixiang, the curator of the Archaeology Department of the National Museum of China, who has strongly supported, encouraged and helped me in many ways since I became a member of the Museum. I also thank Professor Tong Weihua and Professor Zhang Sulin, my colleagues on the archaeological team in Yuanqu, Shanxi, whose work in the Yuanqu Basin has forwarded critical contributions for my study.

I would like to express my thanks to my friends, Fang Dan, Qin Can, Wang Shejiang, Tian Pei, Li Xinwei, Qiao Yu, Ma Xiaolin, Zhang Yongjun, and Sun Zhouyong, and the visiting scholars from China, Zhang Chi, Lei Xingshan and Jiang Leping. I benefited from discussions in archaeology with some of them, and got much help from them in many ways. I spent a lot of pleasant time in Melbourne with them.

I also thank the following fellow graduate students for their help: Anne Ford, Asa Ferrier, Sarah Hayes, Maya Veres, Claudia Zipfel, Jacqui Duncan, Geoff Hewitt, Tom Rymer, Noriaki Sato, and Randy Anderson.

I thank several technological workers of the archaeological workstation in Yuanqu: Wang Wenwu, Yang Hongjun, Lu Zhaoli, and Gao Junxiu, who helped me to survey sites and measure pottery vessels during my fieldwork in Yuanqu.

I am especially grateful to my parents and parents in-law for their encouragement, support and help. My parents have always supported and helped me unselfishly whenever I needed.

Finally, I would like to express my special gratitude to my wife, Li Changxiu, who has beared almost all the housework and looked after our daughter over the past three years. Without her understanding and support, it would have been impossible for me to complete this book.

Chapter One

Introduction

1.1. Research Objective and Background

This study investigates settlement patterns and pottery production from the Neolithic to the early Bronze Age in the Yuanqu Basin, north-central China. The settlement pattern studies are intended to reconstruct the evolutionary process from simple to complex societies in the Yuanqu Basin. Building on this basis, the primary aims of my study are to reveal the changes of modes of pottery production through time, provide insights into the development of specialisation, and examine the relationships between ceramic production and social complexity. Pottery analyses focus on the late Neolithic and the early Bronze Age (ca. 5000-1300 B.C.).

The Yuanqu Basin is located in the south of Shanxi province, just north of the Yellow River, and adjacent to the heartland of the Central Plains (middle Henan Province) where early states emerged. A large amount of archaeological data from this region has been accumulated over the past twenty years, and provides a viable basis for the studies of settlement patterns and craft production.

In the past decades, most Chinese archaeologists have been dedicated to constructing temporal and spatial frameworks and interrelationships among different archaeological cultures. But for the last decade, with the advance of archaeological research and the increase of new discoveries, many archaeologists have changed their interests to research on social politics, economy and ideology. Above all, the origins of Chinese civilisations and the early states have broadly attracted the attention of Chinese archaeologists (Chen 1999; Dai 2001; Liu and Chen 2001a). Meanwhile, more and more scholars attempt to understand the process of the development of Chinese civilisations and social complexity (e.g., Liu 1996, 2004; Yan 1992, 1996, 1999; Zhao 2000).

Settlement pattern studies, or settlement archaeology, has become one of the most popular approaches in Western archaeology since World War II, and has been successfully applied in many regions of the world. This approach is especially effective in studying changes of socio-political organisations and evolutionary processes of complex societies (Sabloff and Ashmore 2001). In China, settlement archaeology may date back to the 1950s, when Banpo, a Neolithic site was excavated on a large scale in order to reveal social relationships and kinship organisations (Institute of Archaeology 1963). But until the 1980s, settlement archaeology remained a relatively minor approach as archaeologists focused on reconstructing social organisations at the community level (e.g., Xi'an Banpo Museum 1988; Yan 1981, 1989a). Since the 1990s, settlement archaeology has become more and more popular in China. Some prehistoric sites have been excavated on a large scale with the aim of examining settlement patterns (e.g., Beijing University and Henan 1998, 2000; Inner Mongolia Team 1997; Institute of Archaeology 2001). Meanwhile, an increasing number of Chinese archaeologists have become interested in regional surveys and interrelationships among settlements, and have conducted surveys and studies of settlement patterns at the regional level (e.g., Beijing University et al. 1992; Dai 2004; Henan Institute of Cultural Relics et al. 1999; Li et al. 1993; Zhang 2003; Zhao 2001). In addition, some archaeologists trained in the Western archaeological tradition have made or are making important contributions to Chinese settlement archaeology (e.g., Liu 1994, 1996, 2000, 2004; Railey 1999; Shelach 1997, 1999; Underhill 1994; Underhill et al. 1998; Underhill et al. 2002). Since the mid-1990s, several Sino-foreign collaborative field projects have been carried out using the full-coverage survey strategy, a kind of systematic regional survey method. These projects are designed to reveal long-term socio-political changes at the regional level, and are bringing the new ideas and methods to Chinese settlement archaeology (Liu and Chen 2001b).

The study of craft production has become another major focus of Western archaeology in the past two decades, and most studies in this field are concerned with social and political structures (Costin 2001). The relationship between modes of production and social patterns is one of the predominant issues in the studies of craft production. A large number of studies have discussed the interrelations among standardisation, specialisation, and social complexity, especially in pottery research (see Rice 1991; Rice 1996). In China, studies of artefacts focus on formal, technological and functional aspects, and few pay specific attention to modes of production and economic organisation. Only in recent years have some archaeologists begun to address the relation between craft production and social patterns. In addition, a few Western archaeologists have explored the relationships between craft production and socio-political patterns in several regions in China (i.e., Flad 2004; Railey 1999; Underhill 1990, 1991, 1996, 2002). In general, research of craft production oriented to political economy is far less popular than that of settlement patterns, and is still in its infancy in Chinese archaeology.

This study is an attempt to combine the two approaches (settlement patterns and craft production) to understand the social process of the Yuanqu Basin.

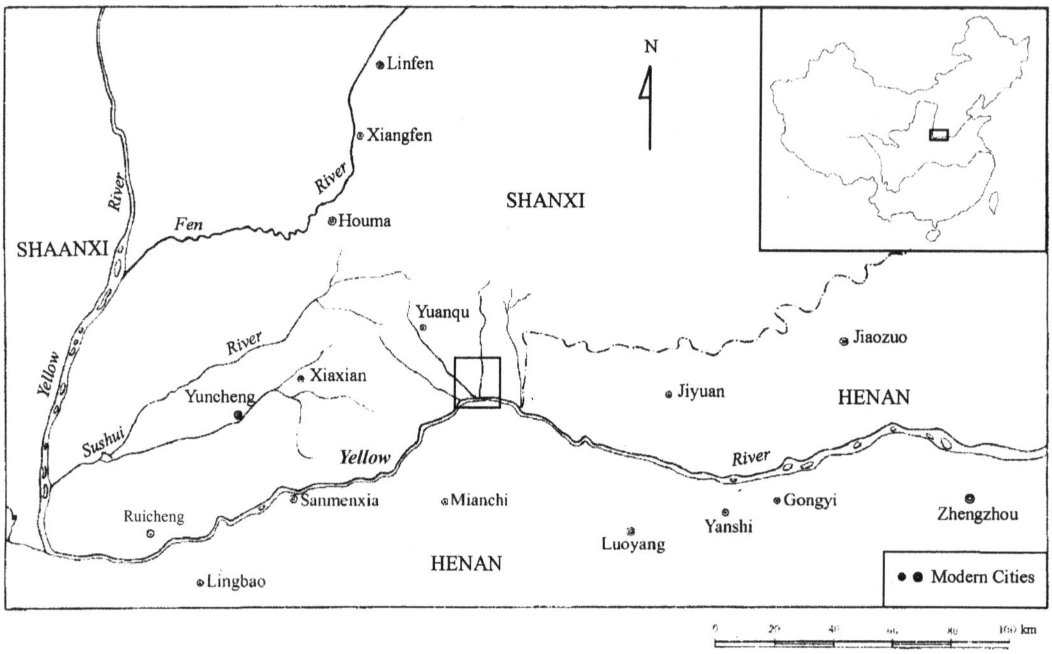

Figure 1-1. Location of the Yuanqu Basin in Shanxi and Henan (The small square indicates the location of the Yuanqu Basin).

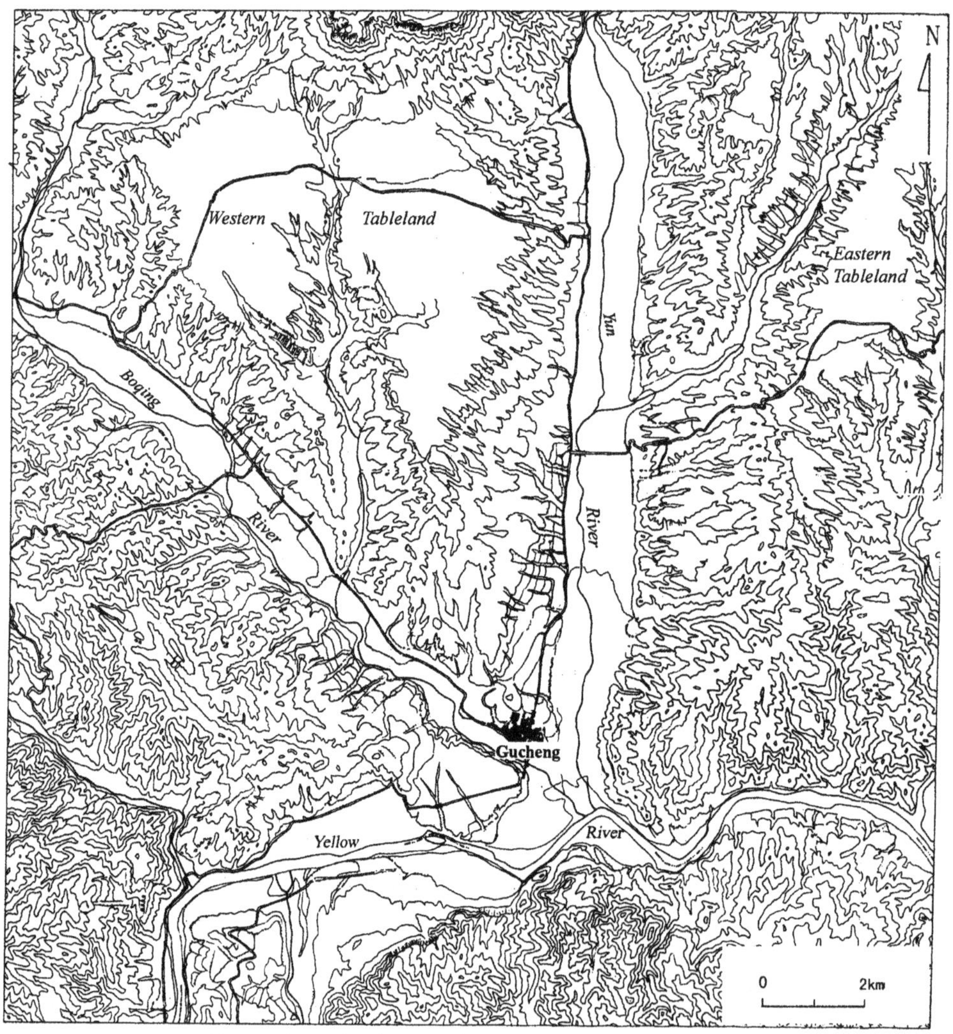

Figure 1-2. Physiographic map of the Yuanqu Basin.

1.2. Location and Natural Environment of the Yuanqu Basin

Yuanqu is one of the southernmost counties in Shanxi Province, located in north-central China and the middle Yellow River valley. The Yuanqu Basin subsumes one part of southern Yuanqu County, and a narrow strip at the northern section of Mianchi County, Henan Province. The Yellow River flows from the west to the east along the southern margin of the Basin, which is the boundary between Shanxi and Henan Province. Most of the area of the Basin is on the northern bank of the Yellow River (Fig. 1-1).

The Yuanqu Basin is situated in a closed mountainous environment, surrounded by the Zhongtiao Range. Apart from the channel of the Yellow River, there are three main routes through mountain valleys which link the Basin and the outside world. One goes north-westward to another larger basin along the Boqing River, a local tributary of the Yellow River, where the Yuanqu city is located. From here, one can go westward to the Yuncheng Basin, or go northward to the Linfen Basin, which are among the larger basins in Shanxi Province. The second pathway is through the valley at the south of the Yellow River to Mianchi County, Henan Province. The third valley extends eastward to Jiyuan County, Henan Province. The last two routes connect with Luoyang and Zhengzhou, which are located at the floodplain of the Yellow River, the heartland of Chinese early states (Fig. 1-1 and 1-2). It is likely that these routes were used from the prehistoric era to the present.

Two local tributaries of the Yellow River, the Boqing River and Yun River, flow through the Yuanqu Basin. The Yun River runs from north to south, and the Boqing River from northwest to southeast, and they debouche into the Yellow River at the south-eastern end of the Basin where they form an alluvial plain. This region and the western floodplain of the Yellow River are the lowest areas in the Basin with an altitude of 205~210 m above sea level (National Museum of Chinese History et al. 1996:4). There are three steps of elevated river terraces along two sides of the Yellow River, Yun River and Boqing River. These terraces are discontinuous and generally blanketed with Pleistocene and/or Holocene loess. The upland between the Yellow River and Boqing River joins with the western mountains and forms a small plateau at the south-central portion of the Basin. In addition, there are two tablelands situated at the west and east of the Yun River, which are called "Western Tableland" and "Eastern Tableland" by local people. In this study the range of the Yuanqu Basin includes these two tablelands (Fig. 1-2). Most archaeological sites and modern villages in the Basin are distributed on the second and third river terraces, the upland, and the two tablelands.

The climate of the Yuanqu Basin is currently characterized by a warm temperate and semi-arid regime, consistent with the climate of most regions of north-central China. Winters are cold and dry, and summers are hot and rainy. The mean annual temperature is 13.2°, and the mean annual rainfall is 640.2 mm (Yuanqu County Chronicles Editorial Board 1993:39-40). Despite relatively low rainfall, the Basin is suitable for cereal agriculture with fertile soils of the loess terraces and abundant water sources provided by the rivers.

Gucheng Town, the administrative centre of the Yuanqu Basin, is located on the alluvium between the Boqing River and Yun River. It was the seat of Yuanqu County for many dynasties until 1959, when the township was moved to the present seat due to the menace of floods (see Fig. 1-1 and 1-2) (National Museum of Chinese History et al. 1996:1-2).

The Yuanqu Basin is situated in the area submerged by the Xiaolangdi Reservoir that was built in recent years and began to fill in 2000. Nowadays most of the Basin has been inundated by water, so its physiographic landscape has changed substantially.

The Yuanqu Basin is confined within an area of 15×15 km. The small and narrow space limits the scale of development of population and polities in the Basin. Moreover, the Basin is circumscribed by rugged mountains, and communication with the outside world is inconvenient. The limitation of the geographic environment played an essential role in shaping socio-political patterns in the Basin (Railey 1999:201-202).

1.3. Archaeological Setting and Chronological Framework for the Yuanqu Basin

From the Neolithic to the early Bronze Age, there were a large number of archaeological cultures in China. Chinese archaeologists define an archaeological culture mainly in terms of artefact assemblages distributed in specific time and space. Above all, pottery is the principal category of artefacts for characterising different archaeological cultures or archaeological periods. On the basis of his theoretical model of *quxi leixing*—the regional system and local cultural series, Su Bingqi suggests that six relatively stable regional divisions existed in prehistoric China (Su and Yin 1981). Each regional system includes a series of archaeological cultures developing successively. These distinct cultural systems interacted with one another and furthered the development of Chinese civilisations. Kwang-chih Chang (1986) and Yan Wenming (1987) suggest similar models. Chang argues that these regional cultures, which had indigenous origins and distinctive styles, became interlinked and formed a "Chinese interaction sphere" (Chang 1986:234-242).

Though advocating the diversity of Chinese prehistoric cultures and origins of Chinese civilisations, Yan also emphasised the leading role of the Central Plain region in the process of the development of Chinese civilisation rather than giving these cultures an equal identity (Yan 1987). The principal reason, according to Yan, is that the Central Plain region is located at the centre surrounded by other peripheral regions, where it

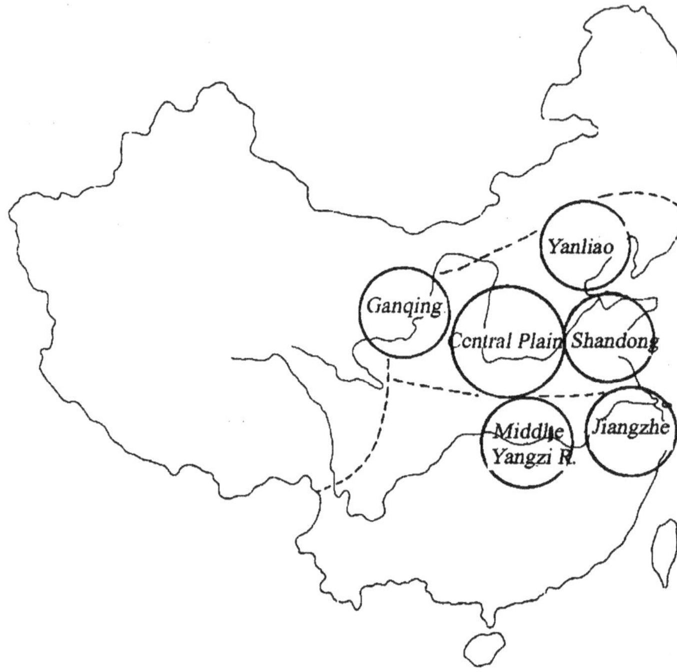

Figure 1-3. Regional divisions of the Neolithic cultures in China (after Yan 1987:fig.1).

could absorb some advanced elements of other cultural regions in the process of interaction with them (Fig. 1-3). A visible fact is that several identifiable early states (Xia, Shang and Zhou dynasties) successively emerged in the Central Plains.

Yan Wenming (1996) divides the Chinese Neolithic into three phases: the Early Neolithic (ca. 10000-7000 B.C.), the Middle Neolithic (ca. 7000-5000 B.C.), and the Late Neolithic (ca. 5000-3000 B.C.). He refers to the time span 3000-2000 B.C. as Chalcolithic. However, copper and bronze artefacts were very few and did not occupy an essential position in social economy during that time, so I prefer to allocate this period to the Late Neolithic. There are few sites dating to the Early Neolithic in the Central Plains. Most of the sites in this phase have been found in southern China. In the Central Plains, the Middle Neolithic includes three synchronic archaeological cultures: Cishan, Peiligang and Laoguantai. The Late Neolithic involves the Yangshao culture (ca. 5000-2900 B.C.), Miaodigou II culture (ca. 2900-2500 B.C.) and subsequent several Longshan cultures (ca. 2500-2000/1900 B.C.) (see Dai 1998a). The early Bronze Age addressed in this study includes the Erlitou culture (ca. 2000/1900-1600/1500 B.C.) and the Erligang culture (ca. 1600-1300 B.C.) (Table 1-1).

It is worth noting that while many Western archaeologists doubt the existence of the Xia Dynasty as the earliest state in China, question the historical connection between Erlitou and Xia, and question whether or not the Erlitou culture represented a state-level polity (e.g., Allan 1984, 1991; Bagley 1999; Keightley 1983; Thorp 1991), most Chinese archaeologists accept Xia and Early Shang as early state-level societies, with Erlitou and Erligang (or Yanshi Shangcheng) as their capitals (e.g., Gao *et al.* 1998, Li 1997a, Wang 1998, Zou 1980). I follow the latter view.

Table 1-1. The Cultural Sequence and Chronological Periodisation for the Central Plain Region.

Age	Phase/Dynasty	Cultures/types & Periods	ca. Dates (B.C.)
Early Bronze Age	Late Shang	Xiaotun (Yinxu)	1300~1100
	Early Shang ?	Erligang Upper (Early, Late) Lower (Early, Late)	1600 ~1300
	Xia ?	Erlitou Late (III, IV) Early (I, II)	2000/1900 ~1600/1500
Neolithic	Late	Longshan Sanliqiao, Taosi, Wangwan III, etc.	2500~ 2000/1900
		Miaodigou II	2900~2500
		Yangshao Late Xiwang, Qinwangzhai, etc. Middle Miaodigou I Early Banpo, Hougang I, etc.	3500~2900 4100~3500 5000~4100
	Middle	Peiligang, Cishan, Laoguantai.	7000~5000
	Early	?	10000~7000

Table 1-2. The Chronological Sequence of Archaeological Cultures for the Yuanqu Basin.

Age	Period	Culture/Type	Chronol. Divisions	ca. Dates B.C.
Early Bronze Age	Early Shang ?	Erligang	Upper Lower	1600 ~1300
	Xia ?	Erlitou	Late Early	2000/1900 ~1600/1500
Late Neolithic	Longshan	Sanliqiao	Early, Late	2500~ 2000/1900
	Miaodigou II	Miaodigou II	Early, Middle, Late	2900~2500
	Late Yangshao	Xiwang	Early, Middle, Late	3500~2900
	Middle Yangshao	Miaodigou I	I, II, III, IV	4100~3500
	Early Yangshao	Dongguan	Early, Late	4600~4100
Middle Neolithic	Peiligang	Peiligang	?	6200~5500

However, at present there is no hard evidence that Erlitou was Xia or that Erligang was Shang. In this study, I avoid referring to Xia and Shang dynasties, but rather simply use the Erlitou state and the Erligang state to refer to the two early state societies.

The Yuanqu Basin is situated in the Central Plain area. The Central Plain area may also be subdivided into a few small local cultural regions (Dai 1998a). The Yuanqu Basin is a component of the south-Shanxi and west-Henan local cultural region (*Jinnan-yuxi Qu*). Table 1-2 presents the chronological sequence of archaeological cultures of the Yuanqu Basin from the Neolithic to the early Bronze Age. It is worth noting that every culture/type (local subclass culture) in this table can be subdivided into different chronological phases (Dai 2002; National Museum of Chinese History et al. 1996, 2001). Some of these will be addressed in the following studies.

1.4. Surveys, Excavations and Studies for the Yuanqu Basin

Formal archaeological work in the Yuanqu Basin began in the 1980s. In 1980 and 1982, The Institute of Archaeology, Chinese Academy of Social Sciences (CASS) surveyed some sites along the Boqing River and Yun River (Shanxi Team 1985), and afterward excavated two Neolithic sites: Longwang'ai (Shanxi Team 1986a) and Fengcun (Shanxi Team 1987).

From the early 1980s, The National Museum of Chinese History commenced large-scale archaeological work in the Yuanqu Basin. Between 1982 and 1986, one work team of the Museum excavated the site of Gucheng Dongguan (hereafter referred to as Dongguan). The report on these excavations has been published (National Museum of Chinese History et al. 2001). In 1984, during the excavation at Dongguan, a walled site of the Erligang culture was discovered at Gucheng Nanguan (hereafter referred to as Nanguan). Recognizing the importance of this site, the Gucheng team conducted investigations and initial excavations at Nuanguan in 1985 and 1986, and then continued the large-scale excavations until 2003 (I joined the excavations from 1993 to 2001). Some results of these investigations and excavations have been reported (National Museum of Chinese History et al. 1996, 1997a, 1997b), but most data remain unpublished. Meanwhile, another work team of the Museum carried out surveys on the south side of the Yellow River, and excavated the site of Bancun from 1991 to 1999, but the data have not been published.

After the middle 1990s, the archaeological work in the Yuanqu Basin started a new phase of salvage excavations with the construction of the Xiaolangdi Reservoir. Several other archaeological teams initiated their excavations at a few sites. The Shanxi Team of the Institute of Archaeology, CASS excavated the site of Xiaozhao in 1993 and 1996, and has published the data (Shanxi Team 1998, 2001). Between 1996 and 1999, the Shanxi Institute of Archaeology carried out excavations at the site of Ningjiapo. The Henan Institute of Cultural Relics excavated the site of Guanjia (Henan) in 1999. But the results of the last two excavations have only been partially reported (Fan 2000; Shanxi Institute of Archaeology 1998).

The primary goals of these investigations and excavations are to establish chronological frameworks, the attributes of different archaeological cultures, the interrelations among these cultures, and the relationships with other contiguous regions (Dai 1998b; Tong 1998; Xu 1996; Zhang 1995). In addition, identifying the layout, configuration and function of the sites was also of concern. One of my studies addresses the settlement patterns and social organizations of Miaodigou I Culture (ca. 4100~3500 B.C.) in the Yuanqu Basin (Dai 2004). In other research, Liu and Chen argue that the principal function of Nanguan, a site of the Early Bronze Age, was to control and transport the resources of copper and salt for the Erlitou and Erligang states—and thereby they relate the Basin to the political economy of the early states (Liu and Chen 2000:53-55; 2002:98-99,111).

In order to reconstruct the process of long-term social changes in the Yuanqu Basin, My colleagues and I (Gucheng Archaeological Team) systematically surveyed most sites on the north side of the Yellow River from 2000 to 2003. Although our surveys were not of full-coverage strategy because of the limitation of time and other conditions (some sites had been submerged by the reservoir water prior to the surveys), they have still furthered our understanding to the changes of settlement and social patterns from Neolithic to Early Bronze Age in the Basin. Although these survey data have not been published, they are completely available for this study.

The first systematic research on the long-term socio-political evolution in the Yuanqu Basin was a doctoral dissertation by Jim A. Railey (1999). He joined the fieldwork in Bancun and surveyed some sites between 1993 and 1995 as a PH.D student at Washington University. Similar to my research, he also focused on settlement patterns and pottery production from the Neolithic to the early Bronze Age. His research on settlement patterns is very detailed, including the analyses of community-level and region-level patterns, but his study on the pottery is relatively cursory. As Railey himself acknowledges, his collection of data is "opportunistic". He could not systematically measure and record ceramics in detail because of the limitation of time and access to the pottery (Railey 1999:242-245). The number, categories and details of ceramics he collected are very limited, so his quantitative and statistical analyses on pottery are restricted considerably. In fact, the settlement data he amassed was not comprehensive. Many sites addressed in my study were not included in his analysis.

According to Railey's research, changes in settlement patterns correspond to changes in social patterns in the Yuanqu Basin. This means that the internal structures and mutual relationships of settlements tended to become more complicated with the development of social complexity from the Neolithic to the Early Bronze Age. Regarding pottery production, Railey drew two major conclusions. First, he saw an increase in the diversity of shape classes over time, consistent with Rice's theoretical model on the relationships between the mode of pottery production and socio-political evolution (Rice 1981, 1984, 1987, 1989). Second, he found little evidence to suggest that pottery production became generally more standardised and specialised with increasing socio-political complexity, contrary to Rice's model (see Chapter Four).

Because of the limitation of data and some analytical methods, some conclusions of Railey's research are problematic and need to be further tested. I have far better data in both settlement and ceramic samples than Railey. Therefore, it is necessary to systematically restudy pottery production and settlement patterns in the Yuanqu Basin. There is ample room for my revisions, though they partially overlap Railey in research. On the other hand, Railey's research contains many valuable elements, and some of his notions are adopted in my study.

1.5. Research Approach, Procedure, and Data

The emphasis of this dissertation is to investigate the change of pottery production in the Yuanqu Basin, while settlement pattern studies provide a socio-political background for research on pottery production. Following a brief synthesis and discussion of relevant theories and methods about social evolution (Chapter Two), I firstly reconstruct the process of social complexity from the Neolithic to the early Bronze Age in the Yuanqu Basin through the analyses of settlement patterns (Chapter Three). The settlement pattern studies include each identifiable cultural period during this time span (ca. 6200-1300 B.C.), but these analyses are comparatively simple due to the limitation in data and the length of thesis. In Chapter Four, I specify and discuss some theoretical and methodological models relevant to the mode of pottery production and specialisation, and detail the methods of collecting ceramic data. The pottery analyses mainly focus on four cultural periods: Early Yangshao (ca. 4600-4100 B.C.), Miaodigou II (ca. 2900-2500 B.C.), Erlitou (ca. 2000/1900-1600/1500 B.C.), and Erligang (ca. 1600-1300 B.C.), which cover the social changes from simple egalitarian society to complex stratified society. Because of the conditions of sites excavated in the Basin, there are abundant ceramic samples available only from these four periods. By measuring and analysing pottery statistically, and combining locations and facilities of production and other evidence, I investigate the development of specialisation and the change of modes of production for the four periods, and discuss the relationships between ceramic production and socio-political forms (Chapters Five~Seven). In the conclusions of this dissertation (Chapter Eight), I summarise the analytical results of this study, evaluate relevant theoretical models on pottery production, and set forth some unresolved problems and unanswered questions for the future research.

A basic expectation for this study is that the Yuanqu Basin experienced a process of social complexity from tribal-level to state-level societies, and correspondingly the pottery production in the Basin became increasingly specialised, and the organisation of production tended to be more and more complex.

The data of settlement patterns subsumes seven sites that have been excavated (Dongguan, Nanguan, Xiaozhao, Fengcun, Ningjiapo, Bancun, and Guanjia), and all other sites that have been surveyed. The main body of ceramic data comes from Dongguan and Nanguan, where I had complete access to the ceramic remains. It is necessary to emphasise that I have better and more adequate ceramic samples for the analyses of pottery production, compared with most similar archaeological studies in other areas of the world.

1.6. Summary

Over the past decade, immense changes have occurred in Chinese archaeology. An increasing number of archaeologists have taken advantage of settlement archaeology to reconstruct the processes of social evolution and explore the origins of Chinese civilisations and early states. The study of craft production is also being extended. My study will contribute to these new research directions by applying current Western archaeological theories and methods to Chinese archaeological data.

A large quantity of data accumulated for the past two decades in the Yuanqu Basin provide a good basis for the diachronic analyses of settlement pattern and pottery production. I will demonstrate the relationships between modes of ceramic production and social forms, which can help to deepen the understanding of social evolution and the formation of early states in central China. I will also test some general theoretical models on pottery production, and answer some questions as these: Whether or not modes of pottery production change along with social forms? What is the relationship between specialised production and social complexity? What are relationships among standardisation, diversity, and specialisation?

Chapter Two

Social Evolution and Settlement Patterns: A Discussion of Theories and Methods

2.1. Introduction

This chapter briefly reviews and assesses some theories and methods related to social evolution (dynamics and process) and settlement pattern studies. The main objective is to establish a set of theoretical and methodological frameworks for the analyses of settlement patterns and social complexity in the coming chapter.

The evolution of human societies has been confirmed by archaeological studies all over the world. Though there still exist various "primitive societies" in some regions, no one can deny the general evolutionary tendency from simple to complex societies for human beings as a whole. "Whether or not sociocultural evolution has taken place is no longer an issue" (Johnson and Earle 2000:2). However, how and why human societies evolve has been controversial over the past decades.

Here, it is necessary to define "social complexity" (or complex society), which is a key conception to understand social evolution. Social complexity can be understood from two dimensions. The first refers to horizontal differentiation in functional specialisation among parts of equivalent rank in a system. The second refers to hierarchical differentiation of organisational components in a social system (Blanton *et al.* 1981:21-22; Hill *et al.* 1996:107-109). In this study the meaning of "social complexity" mainly refers to the second one, as can be seen in chiefly and state societies.

The Yuanqu Basin lies in a peripheral position relative to the acknowledged centres of regional social complexity (such as mid-Henan). Evaluation of general evolutionary theory is not the focus of this thesis. However as the Yuanqu Basin underwent the development from simple to complex societies, the analyses of this process can be placed in the general framework of social evolutionary theory.

2.2. Uni-linear vs. Multi-linear Evolution: Processes of Social Evolution

Following White (1959) and Steward (Steward 1955; Steward and Faron 1959), neo-evolutionary theories have been developed along two divergent pathways: uni-linear and multi-linear evolution. Until the 1980s, uni-linear evolutionary schemes dominated. Their prevalence is mainly attributed to two evolutionary models suggested by Service (1962) based on socio-political organizations (band-tribe-chiefdom-state), and Fried (1967) based on socio-political relationships (egalitarian-ranked-stratified-state). Service's model is more popular and influential in that it seems more explicit, elaborate, and convenient to manipulate. This model has been employed in numerous archaeological and anthropological studies, and evaluated or modified by many archaeologists and anthropologists (e.g., Carneiro 1981; Flannery 1972; Johnson and Earle 2000).

In uni-linear evolutionary theory, "chiefdom" is a key conception and critical evolutionary stage, which bridges the small-scale egalitarian societies and the large-scale hierarchical state societies. According to Service (1962), redistribution is the foremost characteristic of chiefdoms. However, this fundamental chiefly function is questioned by Earle (1977; 1978), and some scholars have redefined the chiefdom in terms of political structure. One of the most representative definitions is offered by Carneiro (1981:45): "A chiefdom is an autonomous political unit comprising a number of villages or communities under the permanent control of a paramount chief ." With the increasing attention given to chiefdoms, there appeared subdivisions of chiefdoms, for example, "group-oriented" and "individualizing" chiefdoms (Renfrew 1974:74), "minimal, typical, and maximal" chiefdoms (Carneiro 1981:47), "simple" and "complex" chiefdoms (Earle 1978; Johnson and Earle 2000; Steponaitis 1978:420; Wright 1984), and so forth. The last category is the most popular and widely applied.

Since the 1980s, uni-linear social evolutionary theory has been increasingly questioned, challenged, and even rejected. The salient concerns of critics focus on the concept of the chiefdom. On the basis of denying the generality and effectiveness of the chiefdom as an evolutionary stage, some scholars criticise and reject the simple typological schemes of uni-linear evolutionism, and emphasise the complexity of social variation and diverse avenues of social evolution (e.g., Blanton et al. 1996; Feinman and Neitzel 1984; White 1995; Yoffee 1993). Just as Blanton and his colleagues state: "The stage typology makes it too easy to think in terms of one repetitive sequence of prototypical forms of societies and to neglect important aspects of variation in sequences and societal types" (Blanton et al. 1999:131). It is especially worth mentioning that Yoffee (1993) suggests a "new social evolutionary theory", which explicitly denotes several distinct evolutionary trajectories based on constraints on kinds of power. In terms of his multi-linear evolutionary model, there exist possible trajectories from "bandishness" respectively to bands, tribes, chiefdoms, or states, and "not all known human societies fall on the progressive steps of a social evolutionary ladder." Meanwhile, his model "does not intend to exclude the possibility of societies in one trajectory (e.g. chiefdom) moving to another trajectory" (Yoffee 1993:72).

While the uni-linear evolutionary models have incurred broad criticism for their weakness of simplicity and narrowness, however, it seems that these critiques cannot completely negate evolutionary taxonomies. Although Feinman et al. (1984) and Blanton et al. (1996) highlight the continuity and multi-dimensionality of social evolution, and Crumley (1995) and White (1995) underscore heterarchy versus hierarchy in complex societies, they do not deny the general evolutionary trajectory from small-scale simple societies, through intermediate societies, to states. Yoffee (1993) presumes diverse evolutionary trajectories, but he cannot adequately demonstrate and exemplify these different evolutionary models. The discussion about social evolution seems to get into such a dilemma: on the one hand, the uni-linear evolutionism or evolutionary typology has been criticised heavily, because they cannot reflect richness and diversity of social evolution; on the other hand, no critic has offered a new acceptable and convincing scheme of multi-linear evolutionism, or a new set of terminologies of social evolutionism. Some basic terms in evolutionary typology, such as "chiefdom", are still widely employed in the discipline with their modified definitions.

As a modification for the band-tribe-chiefdom-state formulation, Johnson and Earle (2000) present a new evolutionary model, which classifies human societies into three types of socio-economic organisations, which actually represent three distinct evolutionary stages:

(1) The Family-Level Group, including the family/camp and the family/hamlet. In family-level societies, the subsistence economy depends heavily on foraging, sometimes in conjunction with the initial horticulture or herding. There could be some flexible and very small-scale camps or hamlets consisting of families, but no clearly demarcated political groups are formed, and social stratification is absent.

(2) The Local Group, including the acephalous local group and the Big Man collectivity. In most cases the subsistence economy of the local group focuses on domesticated species, on the basis of permanent settlements. Local groups are usually subdivided along kinship lines into corporate lineages or clans. Relying on the extent of their common interests, these groups are either acephalous, village-sized units or larger groups integrated by regional networks of exchange headed by Big Man. Political integration is strong within these autonomous local groups, but the power of headmen or Big Man is restricted, and institutionalised inequality does not occur. Service's category "tribe" (1962) covers most local groups.

(3) The Regional Polity, including the chiefdom and the state. The settlement pattern of chiefdoms and states is typically sedentary and hierarchical. Elite is involved in the management of the economy (the subsistence economy and the political economy) in order to get power and political survival. Warfare and territoriality remain central, and the main goal of warfare is to seek expending the political economy by capturing both land and labour and bring them under elite control. The chiefdom can be subdivided into the simple and complex chiefdom. As the regional integration of the chiefdom proceeds, clearly defined offices of leadership emerge at the local and regional levels and are occupied by members of a hereditary elite. Political hierarchy and social stratification thus commonly appear. The development of states and empires involves the extension of political domination, usually by conquest, of a still larger area. Different from the chiefdom, the state develops a set of formal national and regional institutions—an army, a bureaucracy, a law-enforcement system—to handle the increasingly complex functions.

Johnson and Earle's model is a valuable and crucial reference on the social evolution for my study. I still use some basic terms such as tribe and chiefdom to describe different social forms, but they are not used in their original definitions given by Service (1962). When I use "tribe" or "tribal-level society", it equates to "the local group". Likewise, I use "chiefdom" in the same way as Johnson and Earle, which merely refers to the regional polity or the complex society comparatively simpler than the state. I use these terms to describe different social organisations only for comparative and analytical purposes, but this does not mean that I specifically follow a uni-linear evolutionism or the traditional evolutionary typology. I am fully aware of the complexity and diversity of the processes of social evolution in different areas of the world (i.e., the variation and diverse formations of tribes/local groups and chiefdoms). Moreover, in a specific region, the development of social complexity does not necessarily undergo a stereotypical process from the band to the state due to the constraints of the special social or environmental circumstances. This can be verified by the examples of numerous band, tribe and chiefdom level societies in the modern world.

2.3. Mono-causal vs. Multi-variate Explanations: Dynamics of Social Evolution

With respect to the causes of increase in social complexity, it is necessary to distinguish between two different definitions: the origin of institutionalised inequality and the origin of states. Since the chiefdom is the first society of institutionalised inequality, I believe that its origin represents the origin of institutionalised inequality. Some scholars contend that primary factors leading to the rise of chiefdoms and states are the same, so that the rise of states reflects merely the quantitative, scalar change following the emergence of chiefdoms, and the origins of states can be predicted by the rise and development of chiefdoms (Carneiro 1981:38; Earle 1997:14). In most cases, however, they are treated separately. Here, I also consider the origins of chiefdoms and states as two distinct realms and involving different casual factors.

In the past decades, exploring the origin of social inequality has been a crucial concern in archaeology and anthropology, and many explanatory models have appeared. Hayden (2001:244-246) divides most current major models into five groups of basic approaches. These approaches can also be classified into three categories. The first emphasises cognitive, social, and cultural factors (cultural or personal values leading to inequalities). The second stresses material, ecological, and functional factors (population pressure, adaptations to food fluctuation, warfare, control over trade or land, and the like). The third involves "political models": individual self-aggrandisement and self-interest combined with economic surplus production, which enables some community members to control others' labour by using a variety of different strategies including bride wealth, extortion, warfare, rituals, and feasts. Self-interest, surpluses and control over labour are major motivations behind the emergence of socio-economic inequality. Hayden adopts this third approach (Hayden 2001). This kind of multi-variate explanatory model, combining inherent internal causation (self-interest) and external conditions (surpluses and control over labour), can be regarded as more comprehensive than the two others that maintain largely mono-causal explanations. Nonetheless, Hayden cannot adequately elucidate how institutionalised inequality emerged at a specific time and space, in other words, what leads to a production surplus by which aggrandisers can manipulate labour and procure power over others permanently. It is not appropriate to simply ascribe it to technological development. Accordingly, it is still worth exploring the background conditions behind the emergence of institutionalised social inequality.

The rise of states is another issue of concern to archaeologists and anthropologists. There have been several recent systematic reviews and analyses of a variety of theoretical models on the dynamics of state formation (e.g., Blanton et al. 1999:111-132; Flannery 1972:404-407; Renfrew and Bahn 2000:476-477). A number of researchers have been committed to the search for "prime movers" of state origins. Most of them are in the position of material-ecological or functional-adaptation explanations: the rise of states results from the management of large-scale irrigation, internal conflict, warfare, population growth, environmental circumscription, or external trade. All these models have been questioned and criticised. The critics contend that none of them can be applied to all regions as general laws, and nearly all these factors are the results rather than causes behind the emergence of states (see Blanton et al. 1999:111-132; Flannery 1972:404-407). In opposition to foregoing materialists and functionalists, some researchers highlight the predominant role of ideology (religion, art, cosmology, philosophy, and so on) in the process of state formation. This viewpoint is criticised by Flannery (1972:407):

> It is the hierarchical arrangement of members and classes of society which provides the actual integration in states. The critical contribution of state religions and state art styles is to legitimize that hierarchy, to confirm the divine affiliation of those at the top by inducing religious experience.

Therefore, ideology is only a kind of strategy by which elites strengthen their domination in states, but not a prime mover of state formation, though it might play an essential role in the process of the rise of states.

After rejecting a single factor as the dynamic of state origins, Flannery (1972) puts forth his systems theoretical model:

> In a multivariant model, we might see the state evolving through a long process of centralization and segregation, brought about by countless promotions and linearizations, in response not only to stressful socio-environmental conditions but also to stress brought on by internal pathologies (Flannery 1972:414).

In his conception, "promotion and linearization" are evolutionary mechanisms, and "centralization and segregation" are evolutionary processes, both of which are probably universal. He places the prime movers in socio-environmental stresses (population, social circumscription, warfare, irrigation, trade, symbiosis), probably none of which is universal, "but may be specific to particular regions and societies" (Flannery 1972:409). It seems that this kind of multi-variant approach is more productive for elucidating the emergence of states than the simple mono-causal models. However, it is also questionable in some aspects. If one or any combination of several variables (socio-environmental conditions) among subsystems set off a chain reaction of other changes that leads ultimately to the state, it is still necessary to show why this could occur in the first place. At this point, the systems theory encounters the same problem as mono-causal models.

In conclusion, so far there is still no single satisfactory theoretical model for the emergence of institutionalised inequality and states. These complicated issues still remain to be explored. The resolution of dynamics of social evolution depends heavily on a large number of studies of archaeological cases. I do not intend to explore this issue directly in this thesis because of the limitation of data, but I would like to concern it in the study.

2.4. Settlement Archaeology: An Effective Approach to Social Complexity

Since Gordon R. Willey's well-known settlement pattern study in the Virú Valley, Peru (Willey 1953), settlement archaeology has developed to become one of the most important methods in archaeology. It is exceedingly effective in exploring social patterns and changes of social complexity, just as Earle (1984:2) indicates: "most important for studying cultural evolution archaeologically has been a reliance on the settlement patterns."

Indeed, settlement archaeology is a sort of methodology rather than an independent sub-discipline or theoretical paradigm. It can serve diverse academic purposes as an approach through synchronic and diachronic analyses to investigate interrelationships between human societies and environment, access to and utilisation of resources, land use and population, types of subsistence economy and craft production, exchange and trade, beliefs and ritual, socio-political organisations and their changes, and so on. In brief, settlement archaeology provides a powerful analytical framework for studies of socio-politics, economy, and cultures.

Chang (1968:7) postulates that settlement patterns are constituted by two parts: "microstructure" and "macrostructure":

> The cultural and social structure of a settlement we call its "microstructure." The larger cultural and social systems, on the other hand, composed as they are of individual settlements, become "macrostructures." The microstructure is the model for an archaeological community reconstructed on the evidence from an archaeological settlement, whereas the macrostructure is the model for the large sphere of social/cultural activities (beyond those relevant simply to the community) in which members of the community participate, as well as the sphere of cultural and social influences the community imparts to the outside world during a certain time span.

Through these two models, we can reconstruct micro- and macro-social systems within and beyond a community. In contrast to Chang's model, Trigger (1968) suggests a similar but more elaborate analytical formulation. He divides settlement patterns into three analytical levels: 1) "individual buildings", including size, form, method of construction, internal organisation, and the activities taking place in and around the building. 2) "community layout" (settlement layout), including geographical location, number and density of inhabitants per building, internal organisation of the community, the activities within and around the settlement, and the external demarcation of the settlement. 3) "zonal patterns", implying settlement distribution or the spatial relationships between different communities on a regional scale, and including environmental and resource conditions, the overall density and distribution of the settlements, the distribution of loci of specialised activity such as burial sites, population movement, trade, political organization, warfare, and religious and ideational factors (see Liu 1994:43). This model explicitly elucidates the content of settlement archaeology, and its utility can facilitate our practical analyses.

Some archaeologists, who postulate social complexity in terms of socio-political configuration, correlate settlement hierarchy with decision-making hierarchy and socio-political patterns. Usually, simple chiefdoms have one level of decision-making authority or control hierarchy above the level of the local community (two-tiered settlement hierarchy) (Earle 1991:3; Steponaitis 1978:420; Wright 1984:42), and states should have at least three levels of decision-making hierarchy (four-tiered settlement hierarchy) (Wright 1977:389; Wright and Johnson 1975:267). For complex chiefdoms, however, there are some variations among scholars. Steponaitis (1978:420) suggests that complex chiefdoms have two or three tiers of political hierarchy, whereas Wright (1984:42) contends that complex chiefdoms cycle between one and two levels of control hierarchy above the level of the local community. In contrast, Earle (1991:3) explicitly defines the complex chiefdom as a polity which has two levels in the political hierarchy above the local community, which corresponds to a three-tiered settlement hierarchy (Peebles and Kus 1977).

In spite of divergences among scholars, they all agree that there are positive correlations between settlement hierarchy, decision-making hierarchy, and social complexity. These hierarchical models of complex societies provide us a set of tangible criteria to investigate socio-political organisations through regional settlement patterns.

Blanton and his colleagues suggest a more complicated systems model in relation to settlement archaeology to evaluate social evolution (Blanton et al. 1993; Blanton et al. 1982). They are concerned with four "core" factors in explaining changes in human societies, namely *scale, integration, complexity, and boundedness.* Scale "refers to the number of people incorporated into the society and/or the size of the area involved." "Integration refers to the interdependence of units." "Complexity refers to the extent to which there is functional differentiation among societal units." "Boundedness relates to how the population of a social system interacts with other population outside its boundaries, in exchanges of energy, materials, people, genes, and information" (Blanton et al. 1993:14-18). This model avoids the simplicity of settlement hierarchy/political hierarchy models, and enables us to examine socio-political systems with more elaborate and reasonable approaches.

For the assessment of the integration of settlement/social systems, the rank-size distribution of settlements is a useful strategy. The basic form of rank-size distributions, as observed in many different settlement systems, can be defined as this: a settlement of rank **r** in the descending array of settlement sizes has a size equal to **1/r** of the size of the largest settlement in the system. Rank-size distributions can also be illustrated in common logarithms and the result is a straight line called the "log-normal line". The log-normal distribution in accordance with the rank-size rule is regarded as the indicator of well integrated regional settlement systems (Johnson 1981; Liu 1996:240). However, in most cases archaeological inferences are derived from the manner and degree to which rank-size distributions deviate from log-normal. There are two basic types of deviation from rank-size linearity: "primate (or concave)" and "convex" distributions. Primate distributions are those in which the largest settlement is larger than the rank-size rule would predict, or the other settlements in the system are smaller than the rank-size rule would predict. A primate pattern may indicate an extraordinary centralisation of political or economic functions in the social system. Convex distributions are those in which the largest settlement is smaller than expected by the rank-size rule, or the other settlements are larger than expected. A convex pattern indicates a relatively low degree of political and economic integration in a social system. Generally, as they become increasingly integrated, settlement systems will shift from convex to log-normal to primate distribution (Falconer and Savage 1995; Johnson 1977, 1980, 1981, 1987).

Since the rank-size analysis was introduced into archaeology from economic geography, it has been used productively in a number of regional settlement pattern studies (e.g., Adams and Jones 1981; Blanton et al. 1982; Falconer and Savage 1995; Hodder 1979; Johnson 1981, 1987; Liu 1996, 2004; Wright 1986). However, it is noteworthy that rank-size distributions may also be affected by problems of settlement system scale and boundary. Both primate and convex distributions are likely to be produced artificially by articulating different settlement systems or partitioning a single system. Therefore, potential spatial boundary problems involving system pooling or partitioning should be considered (Johnson 1981). It is necessary to define the boundary or range of settlement systems before investigating social integration through rank-size distributions (assessing a single settlement system or multiple settlement systems). In this study, the Yuanqu Basin is a closed and highly circumscribed small basin, so it is relatively easy to identify boundaries of regional settlement systems in most circumstances.

2.5. Summary

This chapter briefly discuss some theoretical and methodological models on social evolution and settlement pattern studies. Obviously, there are many controversies about processes and dynamics of social evolution, and the exploration of these fundamental issues will be the long-term task of archaeology.

I will use some redefined conceptions of social evolution (such as the tribe/local group, chiefdom, and state) to describe the specific social developmental process in the Yuanqu Basin. As a means of facilitating description and cross-cultural comparisons, some form of classification seems inescapable (Railey 1999:99).

The analyses of settlement patterns and social complexity in this study mostly rely on the approaches to settlement archaeology discussed above, including the combination of microstructure and macrostructure on diverse analytical levels, the four elements (scale, integration, complexity, and boundedness) investigating social changes, models of settlement-political hierarchy, and rank-size distribution analysis. These analytical methods are the most effective when used in combination.

Chapter Three

Settlement Patterns and Social Complexity in the Yuanqu Basin

3.1. Introduction

3.1.1. Approaches to Settlement Patterns

The primary objective of this chapter is to discuss the process of social complexity from the Neolithic to the early Bronze Age in the Yuanqu Basin in order to provide relevant socio-political contexts for the following studies on pottery production. On the basis of the theoretical models outlined in the previous chapter, I will try to reconstruct the social form for each period through settlement pattern studies. I do not intend to specially discuss the issue of dynamics of social evolution due to the limitation of data.

In terms of the theoretical approaches discussed in last chapter, I conduct the settlement pattern analyses on three levels: individual features, communities and regions, and assess the development of social complexity from four aspects: scale, integration, complexity, and boundedness.

Although some sites have been excavated in the study area, most of them lack adequate settlement data within communities. This limitation of data makes it difficult to reconstruct social organisation and relationships on the level of households and communities. Hence, settlement analyses have to focus on macro-regional patterns for several periods. The models of settlement-political hierarchy outlined in last chapter can be used as the frameworks for reconstructing social organisation on the regional level. I utilise histograms to show settlement hierarchy, and rank-size distribution to illustrate the integration of social systems. As noted in Chapter Two, these two methods have been proved to be very productive for evaluating the degree of social complexity and integration in a given area.

In this study, I investigate the social scale mainly in the light of the number and size of settlements, and the total occupation area for each cultural period. Though there are many factors influencing settlement space and population size (Hassan 1981:64), the number and size of settlements are seen to correlate positively to the population size in a society (Johnson 1975, 1977; Nordbeck 1971). Boundedness here refers to the political boundary and territory of a society. It is not difficult to establish the limits of the boundaries in most situations in my study, since the Yuanqu Basin is situated in a highly circumscribed environment surrounded by mountains and far away from other areas of settlement convergence. Therefore no special concern is needed for the identification of boundary of each period. However, the interaction of the polities between the Yuanqu Basin and other regions will still be considered whenever relevant.

3.1.2. Procurement and Utilisation of Data

There is a total of 46 sites recorded in the Basin. Thirty-nine of them are distributed on the north side of the Yellow River, and only seven sites are on south of the River (Fig. 3-1, Appendix 1). Although no full-coverage surveys were performed in the Basin, many traditional surveys have been carried out by several archaeological teams in the past two decades, and almost every area suitable for human living has been covered. It may safely be said that there is little possibility that sites, especially larger sites, have been missed. Therefore the result of these surveys should be similar to that of full-coverage strategy. My archaeological team (Gucheng Team) recorded all sites discovered by these archaeological teams.

The data for settlement analyses involve all excavated and surveyed sites. As mentioned in Chapter One, seven sites have been excavated by different archaeological teams in the Yuanqu Basin: Dongguan, Nanguan, Xiaozhao, Fengcun, Ningjiapo, Bancun, and Guanjia (in Henan). The data of the first four sites have been published and constitute the main body for the analyses of microstructure of communities. The results of the last three sites have only been partially reported.

Between 2000 and 2003, my colleagues and I resurveyed most sites on the north side of the Yellow River, and re-identified their dates and sizes. The results are not exactly the same as the former records. In this study, the data of these sites are those based on our surveys, while the data of sites on the south side of the River are based on the records from the Bancun archaeological team.

A number of sites in the Yuanqu Basin contain multiple components of cultural periods (see Appendix 1). Distinct components in the same site usually differ in size and distribution, but in practice it is often difficult to identify the area of every component of most small sites under 10 hectares. Fortunately, this situation has little negative impact on the division of settlement hierarchy because small sites under 10 ha are always in the same site-size group for every period (see the following studies). On the other hand, we attempted to identify the exact area of every component for the large sites and some small sites, and the results are very important for establishing the settlement hierarchy of every period. For most small sites, the total area of the site has been viewed as the area of every component.

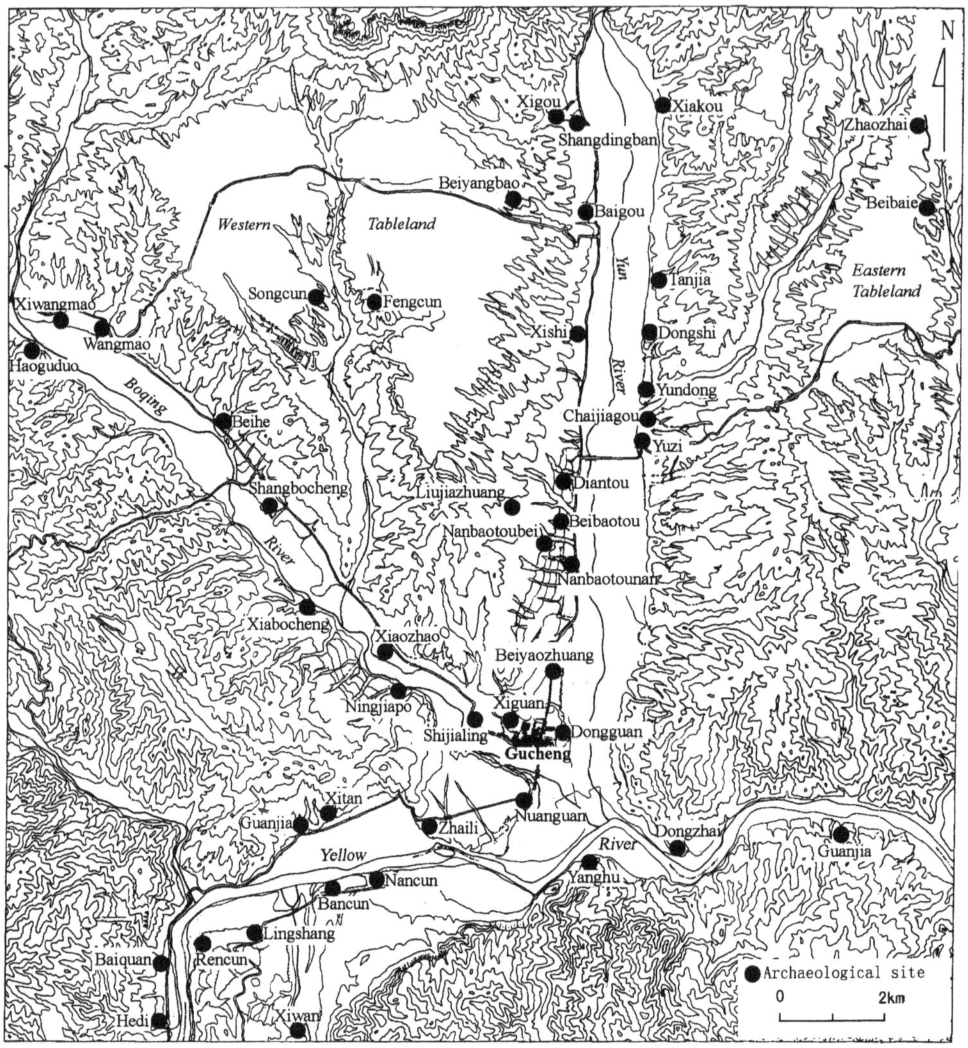

Figure 3-1. Distribution of archaeological sites in the Yuanqu Basin.

As is the case in some other settlement pattern studies, in this study each cultural period is regarded as a relatively stable stage of settlement development. This study involves eight cultural periods (ca. 6200-1300 B.C.): Peiligang, Early Yangshao, Middle Yangshao, Late Yangshao, Miaodigou II, Longshan, Erlitou, and Erligang. The time span of most periods is about 300-500 years. Most sites belonging to one period can be taken to be contemporaneous. Although every cultural period can also be subdivided into different chronological phases, it is impossible to partition all sites into these more precise phases due to the limitations of survey data.

3.2. Peiligang Period

Remains of the Early Neolithic in central China have not yet been discovered, and perhaps they mainly exist in some cave sites like those in Southern China. The Peiligang culture (ca. 6200-5500 B.C.) was one of Middle Neolithic cultures distributed in central China. A number of sedentary Peiligang settlements have been recorded, and some of them have been excavated (Henan Institute of Cultural Relics 1994; Zhao 2001). Archaeological evidence suggests that agriculture had become one of principal subsistence strategies, while gathering and hunting still played an important role in human food production (Henan Institute of Cultural Relics 1999; Zhao 2001).

The core area of the Peiligang culture lies in central Henan Province. The Yuanqu Basin is situated at the northern margin of this cultural region. Only three Peiligang sites have been found in the Basin (Fig. 3-2, Appendix 1), all of them on the south side of the Yellow River. It is reasonable to infer that the earliest Peiligang settlements in the Basin were set up by the immigrants from outside areas, and the Yellow River appears to have become a natural barrier for the diffusion of the Peiligang culture in this area.

Of the three sites, Bancun and Guanjia have been excavated, but the data belonging to this period have not been published. I know, however, that the uncovered features at both sites are mainly ash pits, and there are a

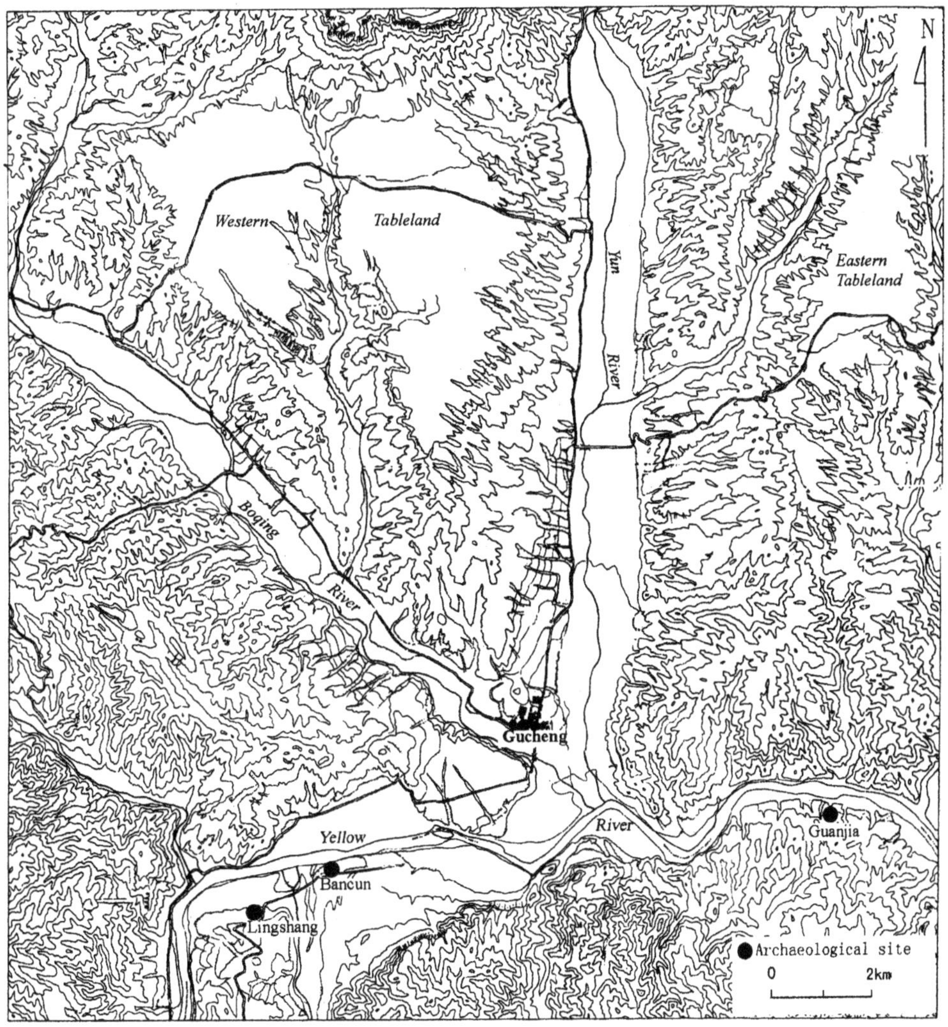

Figure 3-2. Distribution of Peiligang sites in the Yuanqu Basin.

few houses at Bancun. There is therefore not enough information to reconstruct the social organisation within these communities. At the same time, it is not feasible to use the histogram and rank-size curve to manifest settlement hierarchy and social integration for this period because of the limited number of sites. In addition, the size of each site is very small (see Appendix 1). Although the total area of the Guanjia site is 9 ha, the Peiligang remnants are only distributed at the south-eastern part of the site, so the actual area of the Peiligang component at this site is less than 4.5 ha (In the following analyses I will consider 4.5 ha as the hypothetical area of the Peiligang and Erlitou components at the Guanjia site). Accordingly, there was no differentiation of settlement hierarchy for this period in the Basin, and socio-political integration on the regional level had not developed yet.

This conclusion is supported by the studies of Liu (2004:78,128,162-163) and Zhao (2001:52) on the entire Peiligang culture. On the basis of analyses of settlements and cemeteries, both Liu and Zhao argue that the Peiligang society was egalitarian in nature, and organised by kin-groups. Railey reaches similar conclusions. He further contends that Peiligang societies consisted of small-scale, non-stratified, territorial local groups (Railey 1999:147), but argues that based on present evidence Peiligang populations in the Yuanqu Basin were probably organised as autonomous families/hamlets (Railey 1999:582). The hamlet belongs to a kind of family-level groups, whereas the local group consists of many families, and is usually subdivided into corporate lineages or clans (Johnson and Earle 2000:32-33). Some studies based on settlements, especially cemeteries, indicate that lineages and clans existed generally in Peiligang communities (Dai 1996; Zhao 2001; Zhu 1988). Therefore, although all Peiligang settlements are small in the Basin (similar to many sites outside the Basin), the organisations of lineages or clans cannot be ruled out completely. In other words, these autonomous villages in the Basin were possibly organised at the local group level (Johnson and Earle 2000:33).

From the end of the Peiligang to the early Yangshao period in the Yuanqu Basin, there is a lack of archaeological remains. This is a problem that needs to be resolved in the future.

3.3. Early Yangshao Period

The Early Yangshao remains (ca. 4600-4100 B.C.) in the Yuanqu Basin can be subdivided into two distinctive archaeological cultures which are mainly distinguished by ceramic assemblages (National Museum of Chinese History et al. 2001). The earlier one is represented by Dongguan I, which is an indigenous culture distributed in southern Shanxi and western Henan. The later one is represented by Dongguan II, a culture (the Banpo culture) which originated in middle Shaanxi Province, immediately to the west of Shanxi and Henan. This sort of cultural subdivision is common in southern Shanxi and western Henan at this time. It is unlikely that one group of people would have abandoned their own cultural tradition voluntarily to totally accept one from other people. This phenomenon was not simply the result of a change in technology. It probably implies that the people from Shaanxi eliminated or assimilated the aboriginal populations of this region (Dai 1998b).

We identified five Dongguan I sites in the Yuanqu Basin (Fig. 3-3, Appendix 1). These settlements, measuring from 1.5 ha to 8 ha, were distributed sporadically over the landscape, so it is certain that there was not subsistence pressure caused by scarcity of the agricultural land or other resources. The histogram of site size indicates that no settlement hierarchy existed in this period (Fig. 3-4). The rank-size distribution shows a convex curve, suggesting a decentralised settlement system (Fig. 3-5). Therefore, the supra-village organisation had not developed at this period.

The Early Yangshao sites in the Basin, Dongguan and Ningjiapo were excavated on a large scale, but at present only Dongguan has been published entirely (National Museum of Chinese History et al. 2001). The Dongguan site is situated at the east of Gucheng town and the western bank of the Yun River. The total area of the site is approximately 30 ha, and the excavations exposed 2700 m^2. Four excavation areas run from the south to the north of the site (Fig. 3-6). Dongguan I remains are limited to the northern part of the site—Area IV. From our survey, its area is about eight ha, a little less than the 11.2 ha recorded in the excavation report (National Museum of

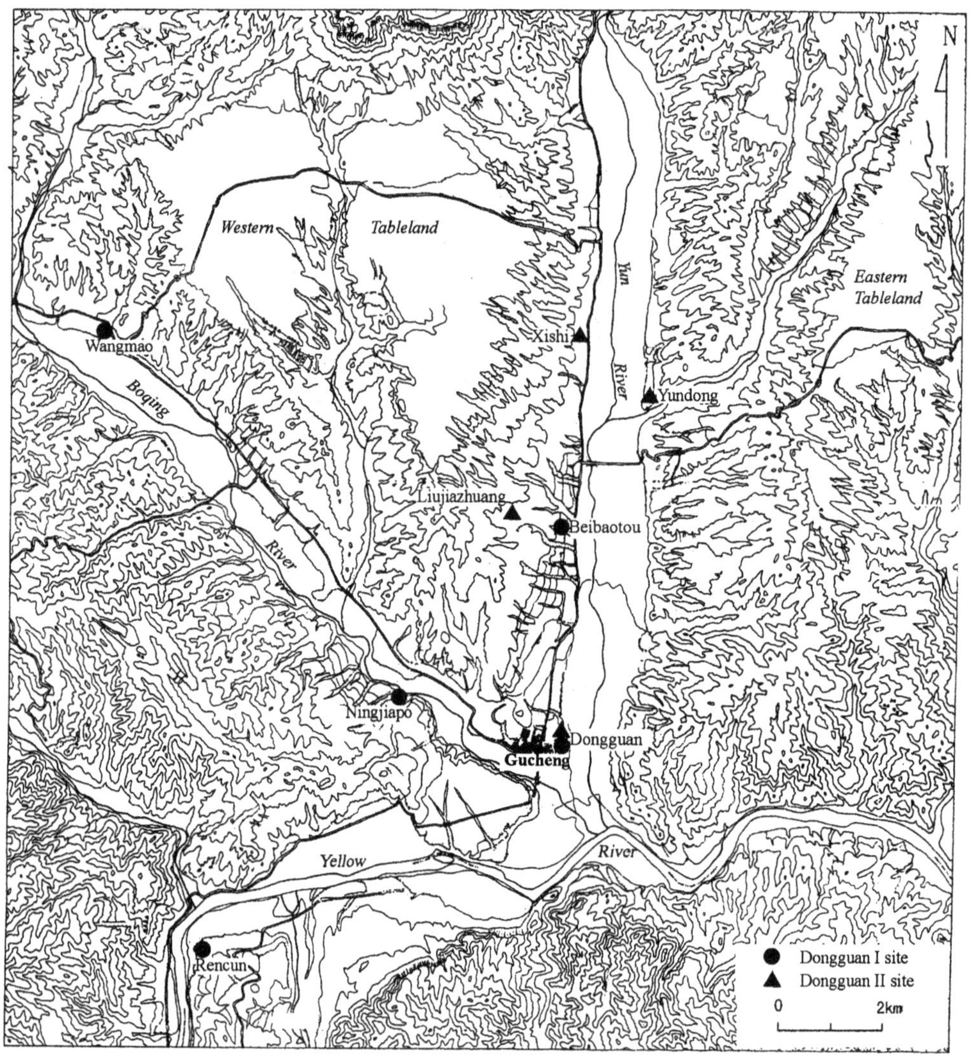

Figure 3-3. Distribution of Early Yangshao sites in the Yuanqu Basin.

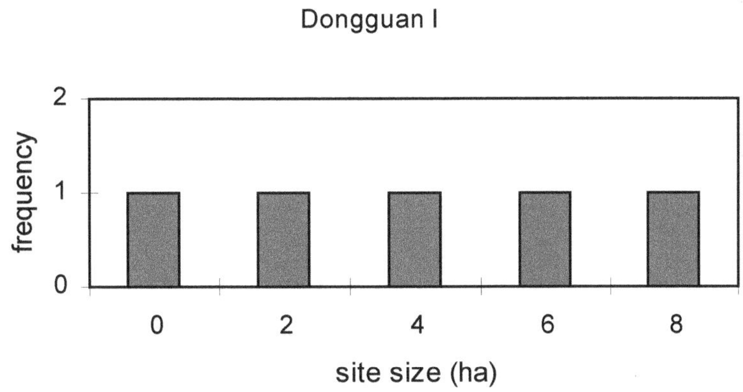

Figure 3-4. Histogram showing the frequency of site sizes of Dongguan I sites.

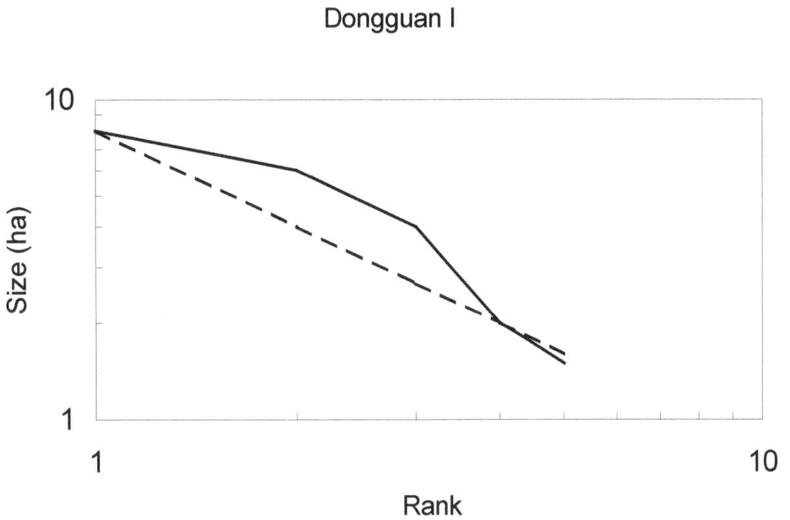

Figure 3-5. Rank-size distribution of Dongguan I sites.

Chinese History et al. 2001:6). The excavated features include pits, kilns, houses, burials, and ditches.

There were 117 ash pits uncovered for the Dongguan I component. The function of most of them is unknown, but we may safely infer that some of them, especially those bell-shaped pits, were storage facilities. Most pits only contained a small number of artefacts, but several of them (e.g., H20, H40, H126 and H132) yielded a large number, with pottery as the main item. However, the social contexts of these storage pits and relations with residential structures in the community are all unknown.

During the investigations in 2001, we found three pottery kilns of Dongguan I exposed along the edges of the eastern bluff in Area IV. Two of them close together at the north, and the other is at the south. The mode of pottery production for Dongguan I will be specially discussed in Chapter Five.

Eight houses of Dongguan I phase were found in Area IV, six of them concentrated in the largest excavation block (Fig. 3-6 and 3-7). All houses are single semi-subterranean structures of oval or round shape. Of these houses, four (F2, F4, F5 and F7) are less than two meters in diameter, and too small to be suitable for the human residences, so they might have functioned as storage or cooking facilities. The other four houses (F1, F3, F6 and F8), ranging from 2.15 to 4.95 m in diameter, might have functioned as residences. The limited excavated area and the small number of uncovered houses do not permit assessment of the distribution, spatial layout, or interrelationships of the houses for this period. Accordingly, we cannot identify the form of social organisations within the community. At present, however, no evidence shows any hierarchical differentiation among the houses.

Only seven burials (M7-M13) were discovered for this phase, and distributed in the same area as those houses (see Fig. 3-7), but they were later than the houses. All are similar in size, with lengths from 1.15 to 1.75 m, and widths from 0.3 to 0.55 m. Apart from a *hu* pot in M9, there are no other grave goods with these burials. Though there are only a few burials, they still manifest certain regularity of distribution: five of them were arranged in two alignments (Fig. 3-7). It may be expected that more graves exist beyond the excavation area, which may either belong to these two alignments or form others. These grave alignments may well have represented

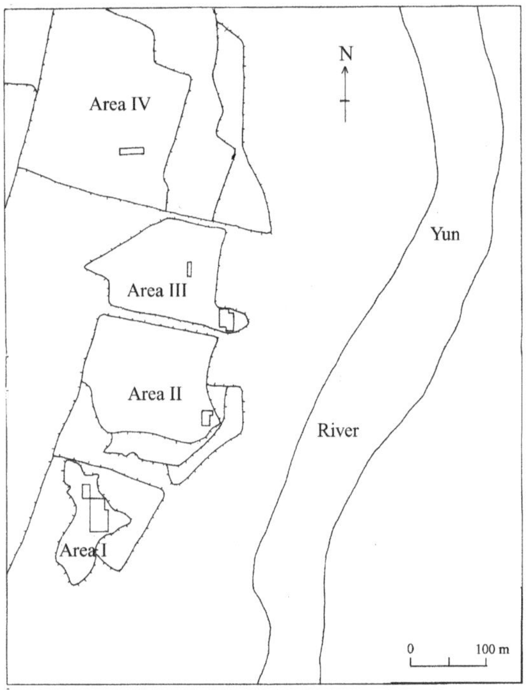

Figure 3-6. Map of the Dongguan site, showing the distribution of excavation areas (redrawn from National Museum of Chinese History et al. 2001:fig.2).

different kin-groups. The uniformity of these burials suggests that there was no burial hierarchy, the relationships among communal members should be egalitarian, and this settlement was probably organised on the basis of kinship.

Several parallel, and even overlapping ditches oriented north-south emerged at the western part of the major excavation block. Among them, G3 is very wide (4.6-4.8 m) and deep (4-4.15 m) (see Fig. 3-7). Although its length and shape are still not clear, its function as a fortification ditch cannot be ruled out, as similar encircling fortification moats existed in many settlements of the Early Yangshao period (Yan 2000).

Generalising the analyses above, especially the macro-regional patterns of settlements and burial patterns, I suggest that during the Dongguan I phase in the Yuanqu Basin, it is very likely that there was no social hierarchy or stratification of inter- and intra-village; every village was autonomous, organised as kin-related corporate groups. The Dongguan I society may have therefore still organised at the level of the local group. These conclusions are similar to Railey's (Railey 1999:584-586). Compared with the Peiligang period, the occupation area of most settlements in this period increased, and the social scale of communities was enlarged.

For the Dongguan II phase, only eight pits were found

Figure 3-7. Distribution of some Dongguan I features in Area IV of Dongguan (redrawn from National Museum of Chinese History et al. 2001:fig.12).

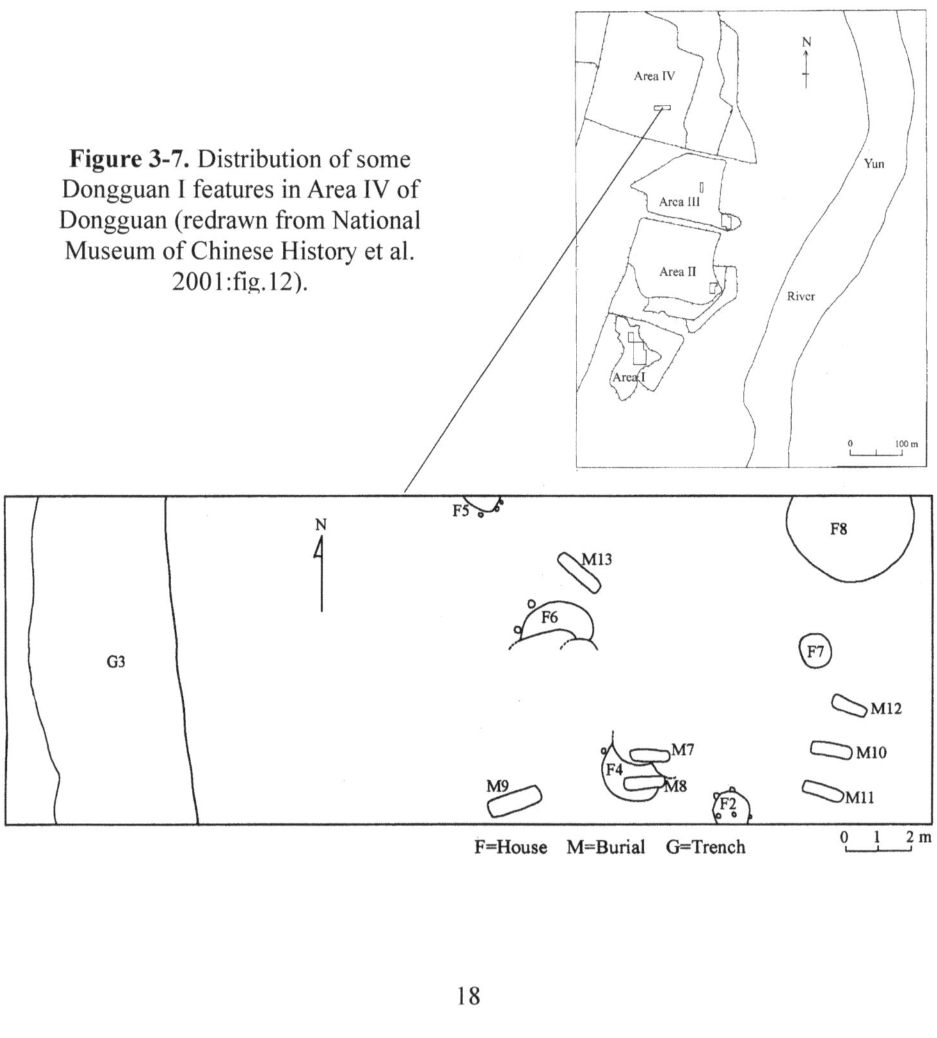

at the Dongguan site. These hardly provide any valuable information in relation to political economy and social organisations. Of these eight pits, one lay in Area I, one in Area II, and the other six pits were distributed closely in Area III. The appearance of two pits in Area I and II was possibly the result of contingent activities, so Area III could be the actual occupation area of Dongguan II component, which is 4.8 ha in size (see Fig. 3-6). Including the Dongguan site, only four Dongguan II sites were found (see Fig. 3-3, Appendix 1). They are all very small, measuring from 0.56 to 4.8 ha. Obviously, no settlement hierarchy is evident for this period. The social pattern of Dongguan II was possibly similar to that of Dongguan I, and the cultural change from Dongguan I to Dongguan II did not give rise to socio-political change. Nonetheless, it is notable that these four settlements were all concentrated in the Yun River valley, and relatively close to one another. This spatial pattern probably implies more intimate relationships among these settlements, possibly suggesting inter-group alliances.

In the Wei River valley in middle Shaanxi Province, several contemporary sites belonging to the Banpo culture (ca. 5000-4000 B.C.) may be helpful in illuminating the social structure of the Early Yangshao settlements in the Yuanqu Basin. These sites include Banpo (Institute of Archaeology 1963), Jiangzhai (Xi'an Banpo Museum et al. 1988) and Beishouling (Institute of Archaeology 1983). All of them have been excavated on a large scale and provide a large quantity of data for settlement and social analyses. The Early Yangshao settlements in these sites were usually divided into several house groups or residential zones, which were enclosed by a moat or by natural barriers. Beyond the residential zones, there were cemeteries corresponding to the house groups. Yan suggests that the pattern of these settlements was cohesive and centripetal, and embodied close social relationships and strong communal consciousness. Every house group may have represented a kin-based group, and they were egalitarian. All settlements were similar in size (about 5 ha for the total area and 2-3 ha for the residential zones), and there was not obvious economic and political differentiation among villages (Yan 1989a:235-236).

The Wei River valley adjoins southern Shanxi and western Henan. The sizes of Early Yangshao sites are similar between the two regions. Furthermore, as stated above, during the Dongguan II phase, it is likely that the people from Shaanxi eliminated or assimilated the indigenous people in southern Shanxi and western Henan. Therefore, it should be comparable for the social structure between them. In fact, the analytical results above also suggest that the social form and relationship of inter- and intra- community are very similar between them.

3.4. Middle Yangshao Period

The Middle Yangshao culture, or the Miaodigou I culture (ca. 4100-3500 B.C.), which developed from the Early Yangshao culture (mainly from the Banpo culture), was extensively distributed in the middle Yellow River. In many regions for this period, especially in middle Shaanxi, southern Shanxi, and western Henan, which are the central areas of the Miaodigou I culture, a settlement hierarchy had commonly emerged (Dai 2004). In some regions, such as the Lingbao region of western Henan, the archaeological evidence suggests that competitive emulation and aggrandising strategy probably existed among high-status families/groups (Ma 2003). An increasing number of archaeologists argue that complex societies may have emerged in the middle Yellow River valley during the Middle Yangshao period (Dai 2004; Liu 2004:164-166; Ma 2003; Yan 1999).

In the Yuanqu Basin, a few remarkable changes occurred in the regional settlement patterns for this period:

(1) The number of sites and the total occupation area increased dramatically, (Appendix 1). Twenty sites were distributed densely over the landscape, the distance between contiguous sites was mostly within two or three km, and even some of them were closely adjacent to each other, such as Nanbaotounan, Beibaotou and Diantou (Fig. 3-8). These situations implied a marked increase of the population and social scale in this period.

(2) The distribution of sites expanded greatly. At the margins of the Basin and near the gorges of the Yellow River several sites appeared, such as Haoguduo, Xiwan, Dongzhai, and Guanjia. Besides, for the first time there was one site (Songcun) located at the highland—the Western Tableland, which was comparatively inconvenient in transportation and far away from the river (see Fig. 3-8). This pattern indicates that the new occupation space and subsistence areas were expanded considerably with the increase of population.

(3) A hierarchy of settlements occurred in the Basin for the first time. The histogram shows a three-tiered settlement hierarchy (Fig. 3-9). The largest site is Beibaotou, measuring 30 ha. The second largest site is Xiaozhao, measuring 15 ha. Other sites are all less than 10 ha in size. The rank-size distribution indicates a near log-normal curve (Fig. 3-10), suggesting a well-integrated social system.

The archaeological studies demonstrate that in a given area the settlement hierarchy usually correlates with the decision-making or administrative hierarchy (Earle 1991; Steponaitis 1978; Wright 1977, 1984; Wright and Johnson 1975). Accordingly, a political hierarchy and supra-village organisation may have emerged in the Basin for this period. Beibaotou, as the largest settlement, might have become the major centre and controlled the whole basin. Xiaozhao might be a minor centre that administered the settlements situated in the Boqing River valley. This situation differed completely from the previous social patterns and probably indicates an emergent complex society for the first time in the Yuanqu Basin. It is likely, therefore, that a chiefdom-level regional polity had been formed.

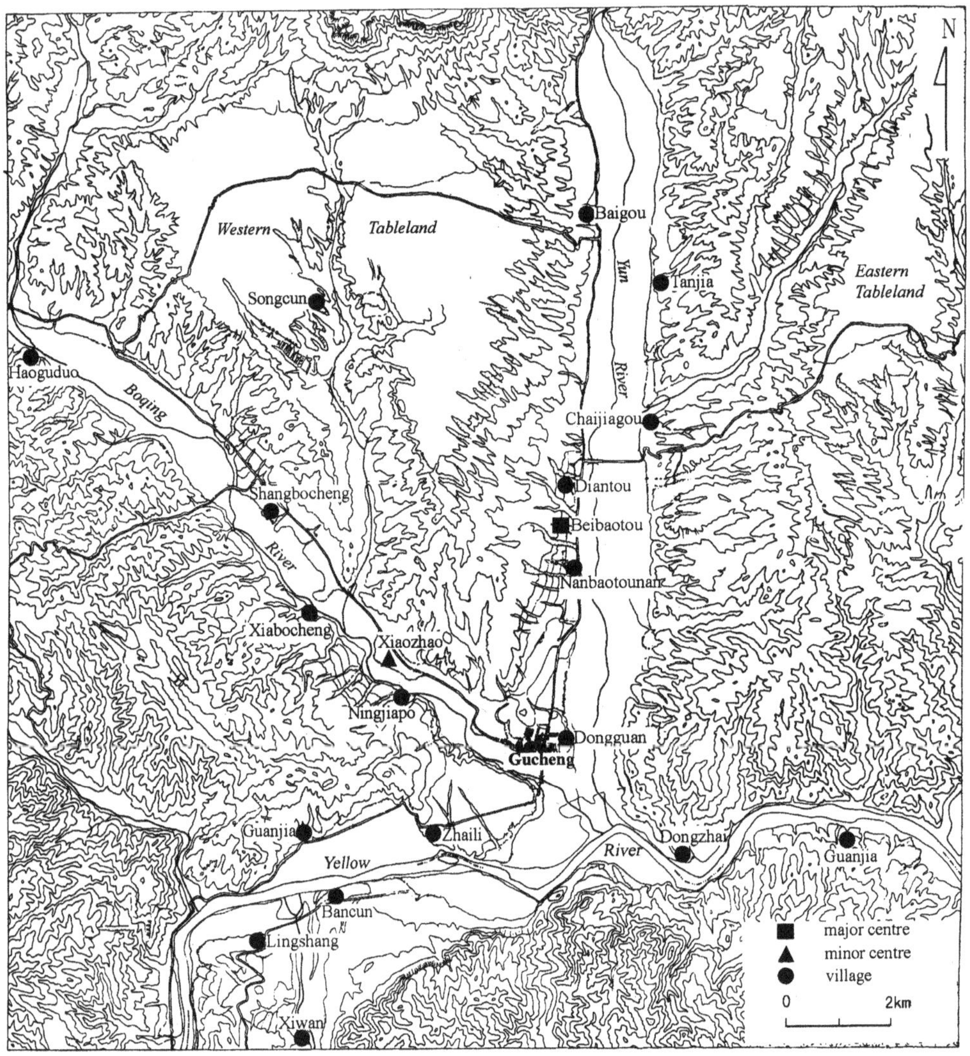

Figure 3-8. Distribution of Middle Yangshao sites in the Yuanqu Basin.

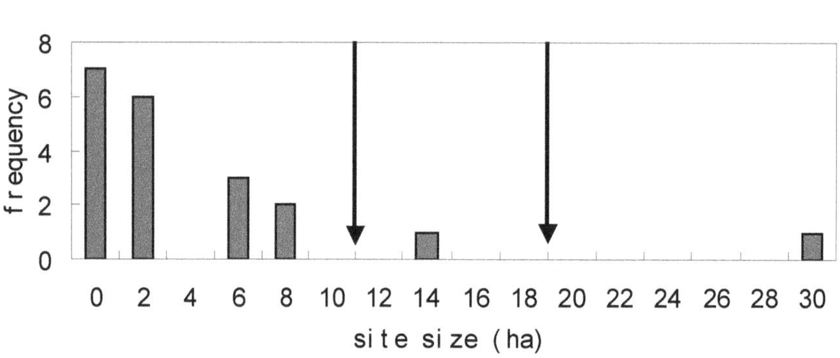

Figure 3-9. Histogram showing three levels of settlement hierarchy in the Yuanqu Basin, Middle Yangshao period.

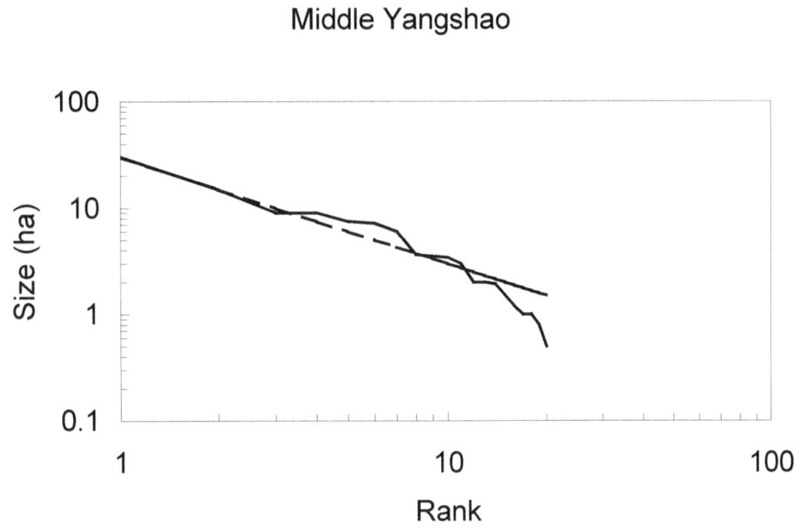

Figure 3-10. Rank-size distribution of Middle Yangshao sites in the Yuanqu Basin.

Nonetheless, at present there is not yet enough information associated with political hierarchy within communities in the Basin.

Of the Middle Yangshao sites in the Basin, Dongguan, Xiaozhao, Ningjiapo, Bancun and Guanjia (Henan) have been excavated. The data of Ningjiapo and Bancun have not been reported. In Dongguan, only one house and seven pits dated to this period were found (National Museum of Chinese History et al. 2001), and they are almost useless for my analysis. At Xiaozhao, the excavated features belonging to the Middle Yangshao component include two houses, eight burials (seven of them are urn burials), four kilns and 26 pits (Shanxi Team 1998, 2001), but the overall layout of this settlement is still unclear, and there is insufficient information to allow us to elucidate its social form and structure. Hence I have to focus on Guanjia (in Henan) for the intra-community analyses.

The Guanjia site (nine ha) is situated at the eastern end of the Basin, on the southern bank of the Yellow River (see Fig. 3-8). Besides a small number of remains dating to Peiligang and Erlitou periods, the excavations here mainly revealed a relatively complete settlement of the Middle Yangshao period. Regrettably, there is no detailed published report available at present. According to a simple report (Fan 2000), however, we can outline the basic structure of this settlement. The Guanjia settlement is encircled by the wide and deep moats at its west and south, and faces the Yellow River with bluffs at its north and east. Apparently, defensibility was a significant concern. Nearly 20 houses lay in the middle and southwest portions of the site. No house hierarchical differentiation was mentioned in the report. Numerous storage pits were distributed around the houses. Most burials were concentrated in the northwest part of the site, and the others were found at the southeast part of the site. Most of them had no grave goods. In addition, more than ten urn burials were uncovered at the southwest of the site, and they were perhaps near the residential sectors. There were more than 50 burials discovered overall, but no evidence indicates burial hierarchy and social stratification. At the northeast of the site, some pits yielded a large number of stones tools, semi-finished stone tools and flakes. Besides, two pottery kilns were also found here. Accordingly, the report speculates that this was a workplace for manufacturing pottery and stone tools.

These records demonstrates that the Miaodigou I settlement at Guanjia had a clear layout and diverse functions. Although we do not know much about it in detail, it may be presumed that this settlement was organised on the basis of kinship, and two residential sectors and two cemeteries possibly represented different and equivalent kin-groups. It seems that there was no obvious evidence of social stratification within the community. The economy of this settlement may have been self-sufficient. Some types of craft products, such as pottery and stone tools, were produced in the settlement. Meantime, the substantial deposits of stone tools and manufacturing debris indicated some kind of specialised production, which possibly took advantage of the availability of stone resource from the nearby river bank and mountains in the gorge.

Guanjia is a small settlement, only measuring nine ha in area. In other regions of the Miaodigou culture, similar to Guanjia, almost all villages and cemeteries lack evidence of marked social hierarchy. However, a few large-scale central places yielded some archaeological evidence probably indicating the emergence of social stratification (Dai 2004). For example, at the Quanhu site (about 60 ha) in Huaxian, Shaanxi province, a burial (M701) yielded a number of grave goods, including a rare and elaborate eagle-shaped tripod (Fig. 3-11), which seems to imply the high-ranking status of the occupant of this burial (Beijing University 2003:74-76; Yan 1999). In the Yuanqu Basin, Beibaotou, the central place, has not been excavated, while the excavated area in Xiaozhao, the secondary largest site, is very limited. The possibility that some remnants are related to social hierarchy in these two sites cannot be precluded.

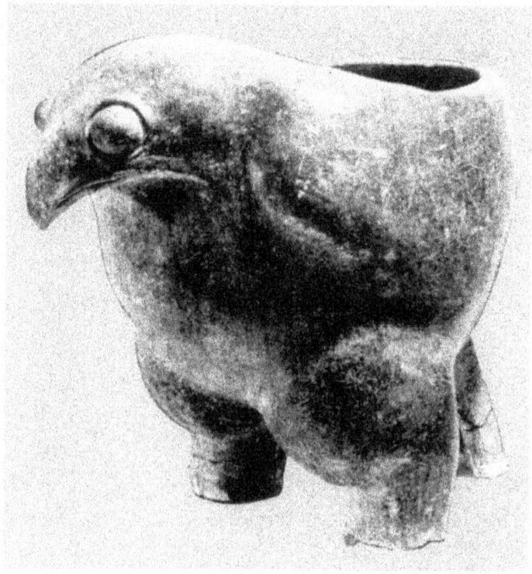

Figure 3-11. An eagle-shaped tripod found at the Quanhu site in Huaxian, Shaanxi (after Beijing University 2003:Photo Page 42).

Zhudingyaun, a region in Lingbao, western Henan, can perhaps provide a good example for understanding the social structure in the Yuanqu Basin. Lingbao is adjacent to southern Shanxi, and Zhudingyuan is only a little more than 100 km southwest of the Yuanqu Basin (see Fig. 1-1). The material remains, especially ceramic assemblages are very similar between the two regions. Zhudingyuan has also been intensively surveyed in recent years (Henan Institute of Cultural Relics et al. 1999; Institute of Archaeology 1999a). Similar to Yuanqu, 19 sites were clustered in this region, also forming three tiers of settlement hierarchy (Liu 2004:165; Ma 2003:40). Differently, Beiyangping, the major centre (90 ha in area), and Xipo, the secondary centre (40 ha) at Zhudingyuan, are much larger than Beibaotou and Xiaozhao in Yuanqu. This situation suggests a larger social scale and population size, which may have something to do with larger arable land available in Zhudingyuan.

Xipo has been excavated in recent years (Henan 1st Team and Henan Institute of Cultural Relics 2001; Henan Institute of Cultural Relics et al. 1999, 2002, 2003). A few large buildings, measuring about 100 m^2 or more, have been unearthed or investigated. The largest one is F105, a square-shaped subterranean structure with a wing corridor around the main building. It measures 516 m^2 in total area, and has a rammed-earth foundation up to 2.75 m in depth (Fig. 3-12). Its construction required a large number of labourers, who were perhaps mobilised from several villages. No domestic refuse was found near this building. Therefore, F105 was very likely to be a large monumental structure, built for public activities rather than for domestic purposes (Liu 2004:84; Ma 2003:85). This monumental structure or ritual ceremony centre may have played a critical role in the regional social integration. The other discovered buildings may have been domestic architecture, indicated by numerous artefacts and faunal remains in nearby ash pits. The faunal remains were predominantly from pig bones, which were very abundant, but meanwhile were of imbalance in spatial distribution. The consumption of pigs in Xipo was thus very likely to be associated with competitive feasting (Ma 2003:208-209, 215, 218). Besides, two buildings yielded grinding slabs used for processing red pigment (cinnabar), which was regarded as a scarce ritual substance. The control over the access to cinnabar and production may indicate that some individuals or families were more successful in gaining higher prestige and power (Liu 2004:85; Ma 2003:90-91, 218). In summary, all evidence above, including feasting behaviour and restricted access to valuables, suggests factional competition for high prestige and power among groups in the Xipo community. These new changes in intra-settlement relationships and the emergence of regional settlement hierarchy, strongly indicate that competitive emulation and aggrandising strategy characterised the trans-egalitarian society at Zhudingyuan with an emergent social hierarchy (Liu 2004:85, 166).

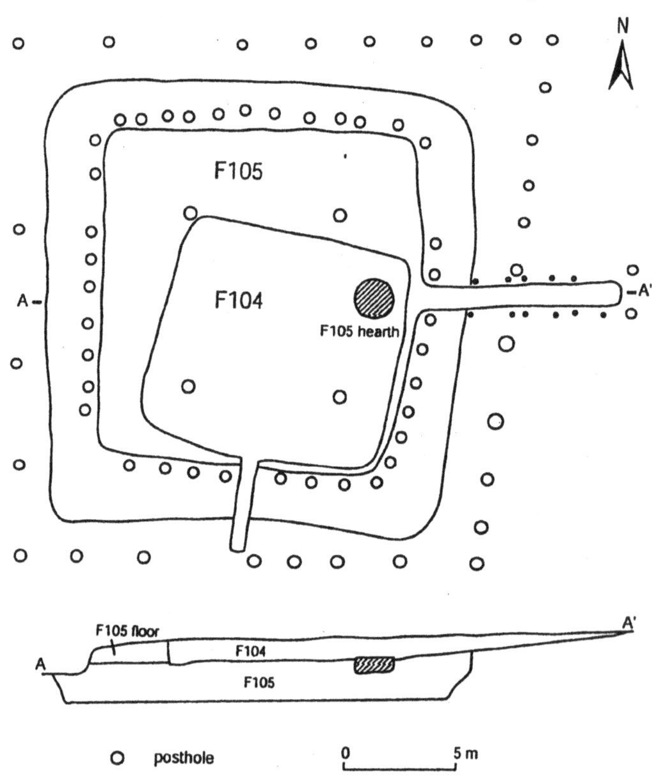

Figure 3-12. A large building (F105) discovered at Xipo in Lingbao, Henan (after Henan Institute of Cultural Relics et al. 2003:fig.3, cited from Liu 2004:fig.4.7).

Like Zhudingyuan, the Yuanqu Basin was also situated in the central area of the Miaodigou I culture, and had a similar regional settlement structure. It is likely that the large public architecture and remnants associated with competitive emulation and aggrandising strategy for prestige and power also existed at Beibaotou and Xiaozhao, the central places in the Basin. Some evidence for this has been found at Xiaozhao: an elaborate house foundation at the southeast section of the site has a layer of wooden floor under the upper lime and grey-clay floors (Shanxi Team 2001:194-196). This sort of special structure was rare among contemporary houses, and was probably related to a special function or high-status family. Therefore, it is reasonable to conclude that the Yuanqu Basin experienced its earliest development of complex society and formed a regional polity during this period.

This argument is different from that of Railey. Due to the limited data available to him, Railey postulated that the clear evidence for social stratification or settlement hierarchies was absent in the Miaodigou I period, so supra-village integration involved the formation of alliances between otherwise villages rather than true, multi-village polities (Railey 1999:587).

Three tiers of settlement hierarchy are usually believed to correspond to two levels of control hierarchy, which are regarded as an indicator of complex chiefdoms (Earle 1991:3; Steponaitis 1978:420; Wright 1984:42-43). According to this proposition, a complex chiefdom could be developed in the Yuanqu Basin and Zhudingyuan during the Middle Yangshao period. However, so far there is not yet sufficient archaeological evidence within communities to support this argument. Because of the limited excavation at the primary and secondary centres of the two regions, it is unclear if there existed segregated elite residences or burials, which are considered as the common traits of complex chiefdoms (Wright 1984:44). At Xipo, all artefacts discovered from buildings (including high-ranking structures) and nearby deposits are utilitarian goods, and no prestige goods were found (Henan 1st Team and Henan Institute of Cultural Relics 2001; Henan Institute of Cultural Relics et al. 1999, 2002, 2003). Though the procurement and use of cinnabar was possibly restricted to elite families/groups, it was more likely to be ritual substance rather than personal possessions. It is possible that the most salient physical attribute in the Miaodigou I societies was the presence of large public facilities rather than personal wealth. After all, the Miaodigou I societies probably were only the beginning of social complexity, so the development of social stratification and political hierarchy may have been limited. Perhaps, such societies may be described as "group-oriented chiefdoms", in which personal wealth in terms of valuable possessions was not important, but the solidarity of the social unit was expressed most effectively in communal or group activities (Renfrew 1974:74).

3.5. Late Yangshao Period

The Late Yangshao culture (ca. 3500-2900 B.C.), which occupied most areas of the Middle Yellow River, was the successor of the Middle Yangshao culture. It subsumed a few local cultural types (subclass cultures) that had distinct ceramic assemblages. The ceramic assemblage existing in southern Shanxi and western Henan, including the Yuanqu Basin, is called the *Xiwang Type*. The Central Plain regions, including middle Shaanxi (the Wei River valley), southern Shanxi, and western & middle Henan (see Fig. 1-3), were the central areas of the Late Yangshao culture (Dai 1998a).

Beyond the Central Plains, especially in eastern China, including the Yanliao region in north-eastern China (the Late Hongshan culture), the Shandong region in the Lower Yellow River (the Middle Dawenkou culture), and the Jiangzhe region in the Lower Yangzi River (the Early Liangzhu culture) (see Fig. 1-3), social stratification and political hierarchy had become fairly remarkable during this period, indicated by high-value prestige goods, high-ranking elite burials and segregated elite cemeteries (Yan 1992, 1999). By contrast, there is little such evidence discovered in the Central Plains. Similar to the previous period, the most salient features were still large-scale public architecture rather than personal wealthy possessions in the Late Yangshao culture. For example, Dadiwan, a central place located in Qin'an, eastern Gansu (which adjoins western Shaanxi), measuring 110 ha, yielded a palace-like large structure (F901) in the centre of the site (Gansu Cultural Relics Working Team 1986; Gansu Provincial Museum 1983). Judging from its size, configuration, and artefacts in the room, this building was very likely to have functioned as a monumental structure for public activities, perhaps including feasting, redistribution, ritual performance, and public gathering, and Dadiwan may have been a major centre in the regional settlement system (Liu 2004:86-88; Yan 1989a:218, 239). Xishan, a small site (3.45 ha), lies in northwest of Zhengzhou, Henan (see Fig. 1-1). Its Late Yangshao component was enclosed by rammed-earth walls, making it one of the earliest walled sites discovered in prehistoric China (Archaeological Team Leader Training Program 1999). Its construction undoubtedly needed a large number of labours who were likely to be assembled from many villages, and may have indicated the existence of the supra-village organisation. Although no high-ranking burials or prestige goods were discovered, the regional social integration showed by these two sites still indicates the development of complex social relationships in some Late Yangshao societies.

During this period, some essential changes of regional settlement patterns occurred in the Yuanqu Basin:

(1) The number of sites substantially decreased, from 20 to 11. The total occupation area correspondingly dropped too (Appendix 1).

(2) The distribution of these sites over the landscape was comparatively clustered. Most of them were distributed around the centre of the Basin—Gucheng, where the largest site, Dongguan, was located. No sites

have been found on the margins of the Basin near the gorges of the Yellow River (Fig. 3-13).

(3) The central place shifted from Beibaotou to Dongguan during this period. The area of the Late Yangshao component at Beibaotou dropped abruptly to 8.4 ha. Dongguan became the largest settlement, measuring 14 ha, but it was far smaller than the central settlement of the previous period.

(4) The histogram indicates two tiers of settlement hierarchy (Fig. 3-14). Dongguan was in the position of the central place. The rank-size curve is slightly convex but near log-normal (Fig. 3-15). This pattern suggests a weakly integrated polity. In other words, though Dongguan may have controlled the whole basin, it probably faced the competition and challenge from other relatively large settlements—such as Beibaotou (8.4 ha) and Xiaozhao (8 ha).

Of the sites that have been excavated in the Yuanqu Basin, four contain the remains of the Late Yangshao period: Dongguan (National Museum of Chinese History et al. 2001), Nuanguan (National Museum of Chinese History et al. 1996), Xiaozhao (Shanxi Team 1998), and Fengcun (Shanxi Team 1987). The majority of features uncovered in these sites are ash pits. For most of them, their function and context in relation to residential structures is unclear, so it is impossible to reconstruct the social structure and political economy in the communities. Apart from these pits, there are only four structures discovered at the Nuanguan site. They are all very small, measuring less than 4 m^2, so some of them are unsuitable for the human residence. In a word, we know little about the social structure and organisations within communities merely based on these data.

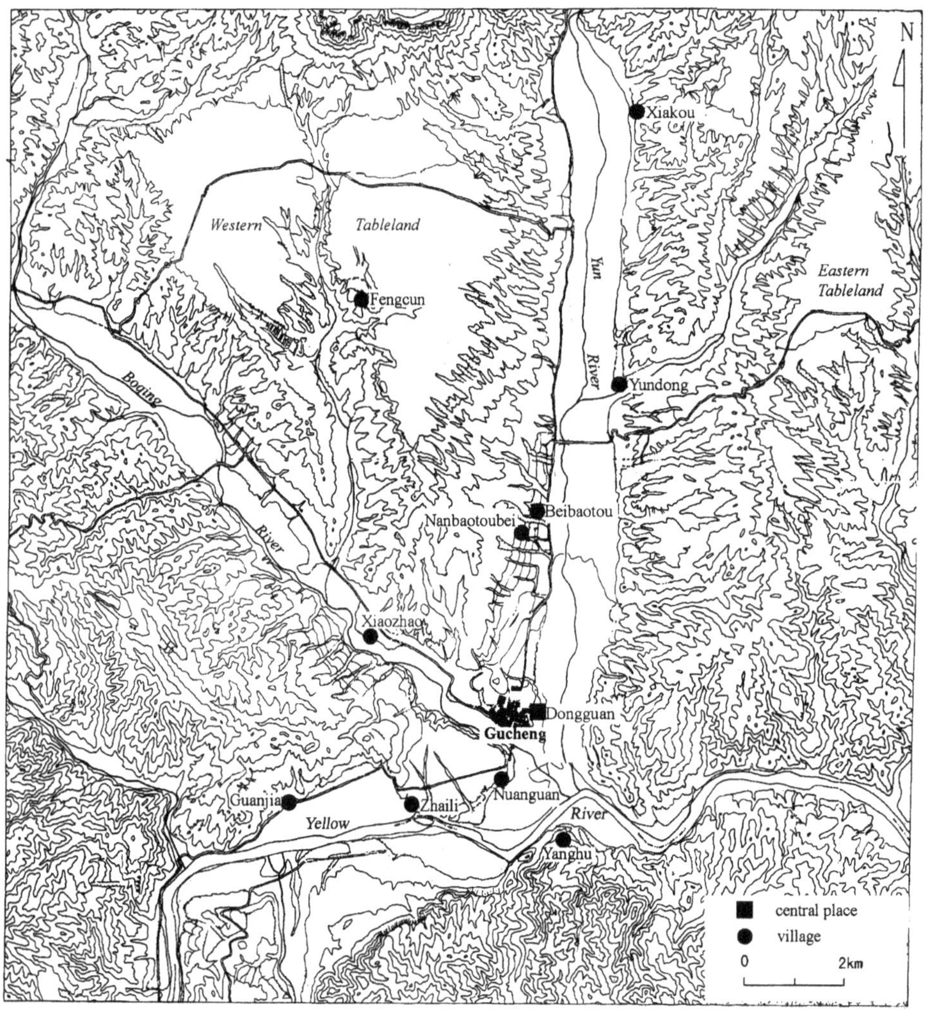

Figure 3-13. Distribution of Late Yangshao sites in the Yuanqu Basin.

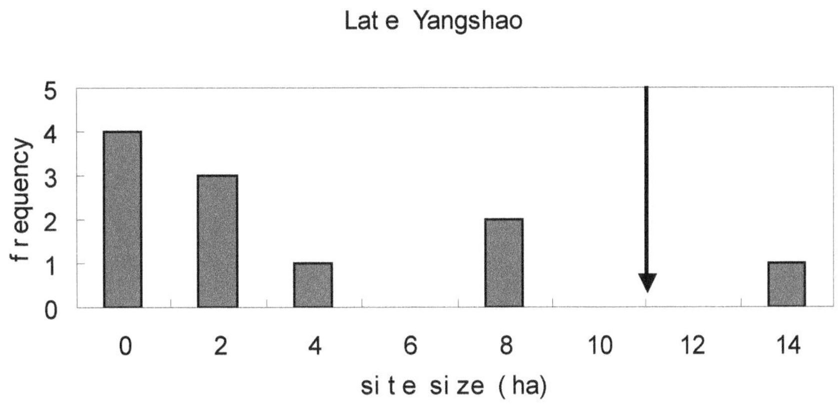

Figure 3-14. Histogram showing two levels of settlement hierarchy in the Yuanqu Basin, Late Yangshao period.

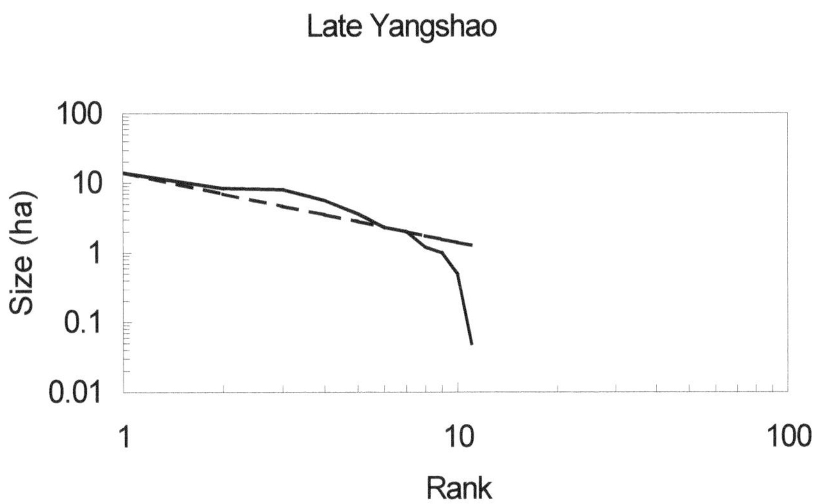

Figure 3-15. Rank-size distribution of Late Yangshao sites in the Yuanqu Basin.

The decrease in the number of sites, the central settlement size, and the total occupation area suggest a decline in population and social scale, while the reduction in the settlement hierarchical level and in the degree of settlement integration may imply a decline in political hierarchy and social complexity in the Yuanqu Basin. According to current archaeological data, it seems that the decline in population, social scale and social complexity was a common phenomenon in the central areas of the Late Yangshao culture, especially in southern Shanxi and western Henan. For example, similar to the Yuanqu Basin, Zhudingyuan, in Lingbao, western Henan, also experienced a sharp decrease in the number of sites (from 19 to 8), the area of central settlement (from 90 to 30 ha), population size, and the settlement hierarchy (from 3 to 2) during the Late Yangshao period (Ma 2003:43). In contrast with some areas, such as the Dadiwan and Xishan cases discussed above, the Yuanqu Basin and Zhudingyuan may have experienced a social decline, and social changes perhaps happened unevenly among different regions during the Late Yangshao period.

However, the two-tiered settlement hierarchy, suggesting one-level decision-making hierarchy above the level of the local community, is usually regarded as a indicator of simple chiefdom (Earle 1991; Steponaitis 1978; Wright 1984). At least, it is likely that such regional settlement pattern still indicates a complex rather than a simple society in the Yuanqu Basin.

In general, based on current data, we know little about social structure and degrees of social complexity in the Yuanqu Basin. I can only infer that during the Late Yangshao period, the Yuanqu Basin may have witnessed a complex regional polity, rather than reversed to a simple society consisting of autonomous villages. Nonetheless, I cannot yet specify the particular social organisation and political structure for this polity.

3.6. Miaodigou II Period

The Miaodigou II culture (ca. 2900-2500 B.C.) is one of the major archaeological cultures distributed in the Middle Yellow River during this period, and mainly stemmed from the Late Yangshao culture, especially from the *Xiwang Type*. Middle Shaanxi, southern Shanxi and western Henan were still the central areas of the Miaodigou II culture (Dai 1998a).

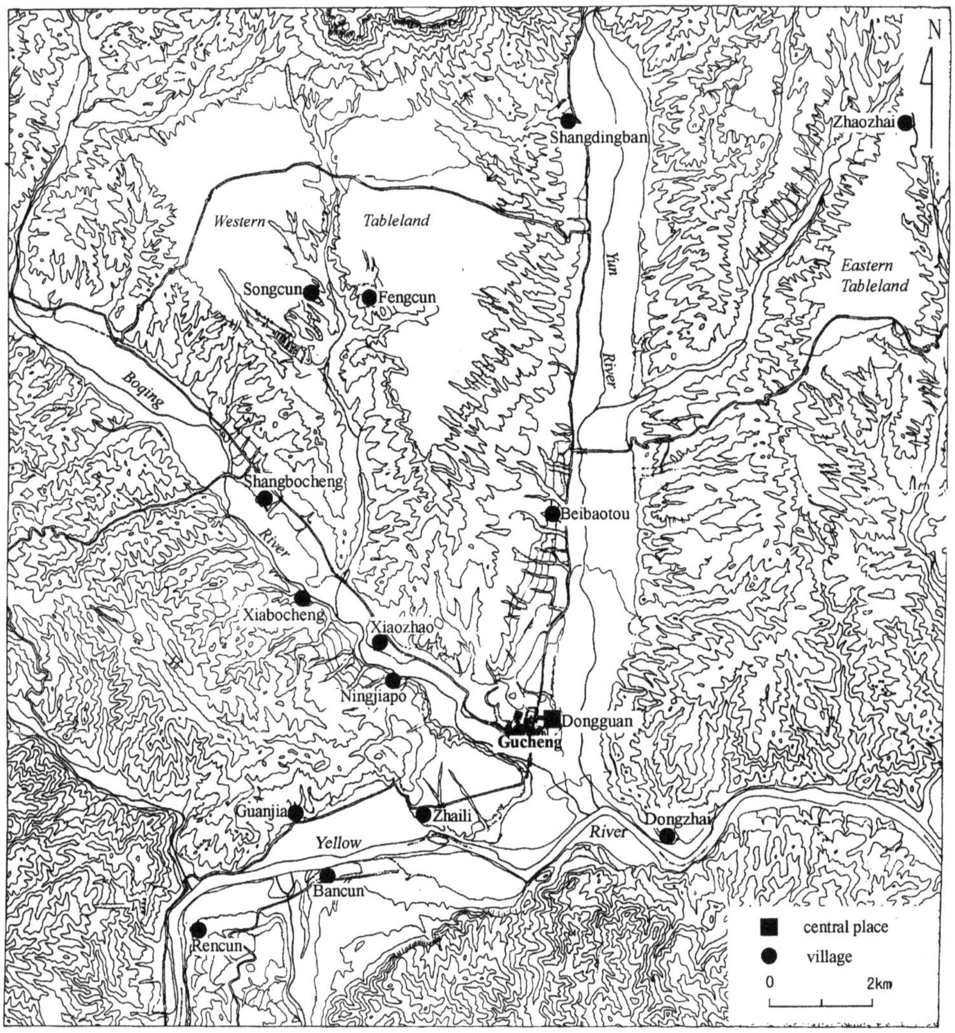

Figure 3-16. Distribution of Miaodigou II sites in the Yuanqu Basin.

Beyond the Central Plains, social complexity further developed during this period, especially in the Lower Yellow River (the Late Dawenkou culture), the Middle Yangzi River (the Qujialing culture) and the Lower Yangzi River (the Late Liangzhu culture) (Yan 1999). Different from the previous period, the archaeological evidence of social stratification also markedly appeared in the Middle Yellow River, even in the Central Plain. For example, in the Linfen Basin that is situated at the north of the Yuncheng and Yuanqu basins in southern Shanxi (see Fig. 1-1), a large-scale central place emerged in the Taosi site, Xiangfen County (Gao et al. 1983; He and Yan 2002; He et al. 2003; Shanxi Team 1983, 1986b; Shanxi Team and Bureau 2003). During the early phase of this site (dating to the Late Miaodigou II period), a highly complex society formed at Taosi as indicated by a rammed-earth enclosure (measuring 56 ha in area), palatial and elite residential zones, large elite burials, and a great number of highly valuable prestige goods, including elaborate painted pottery, wooden artefacts, and jade objects. At the Qingliangsi site, Ruicheng County, which is situated at the north bank of the Yellow River like Yuanqu (see Fig. 1-1), some elite tombs also yielded many elaborate jade items (Shanxi Institute of Archaeology and Museum 2002; Xue 2004, 2005). Jade items were the most popular prestige goods in the Late Neolithic of China. Therefore, not only large monumental architecture but also personal wealth characterised some stratified societies in the Central Plains during the Miaodigou II period. This situation perhaps implies the transformation from simple or group-oriented chiefdoms to complex or individualising chiefdoms in some regions (Renfrew 1974; Wright 1984).

In the Yuanqu Basin, the regional settlement patterns for this period displayed the following features:

(1) The number of sites increased slightly, from 11 to 15, while the total occupation area rose markedly (Appendix 1).

(2) The overall distribution of the sites expanded once again. Besides the Western Tableland, one site (Zhaozhai) appeared on the Eastern Tableland for the first time. These sites were widely distributed over the landscape. Comparatively, the sites along the Boqing River were relatively dense (Fig. 3-16).

(3) Dongguan was still the largest settlement in the Basin, and its area rose to 20 ha. The histogram shows

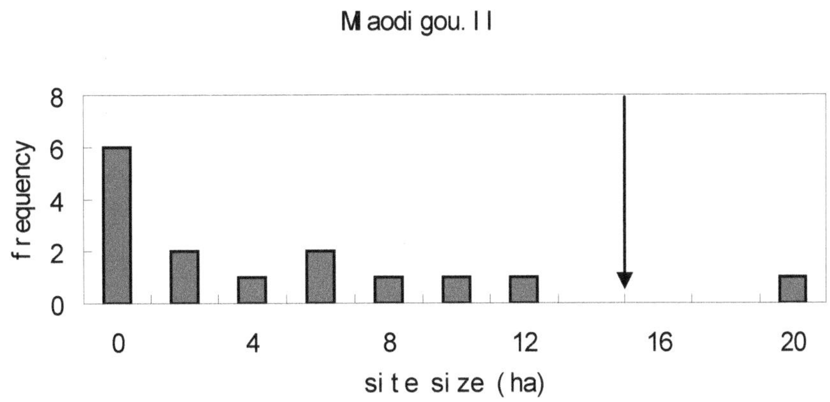

Figure 3-17. Histogram showing two levels of settlement hierarchy in the Yuanqu Basin, Miaodigou II period.

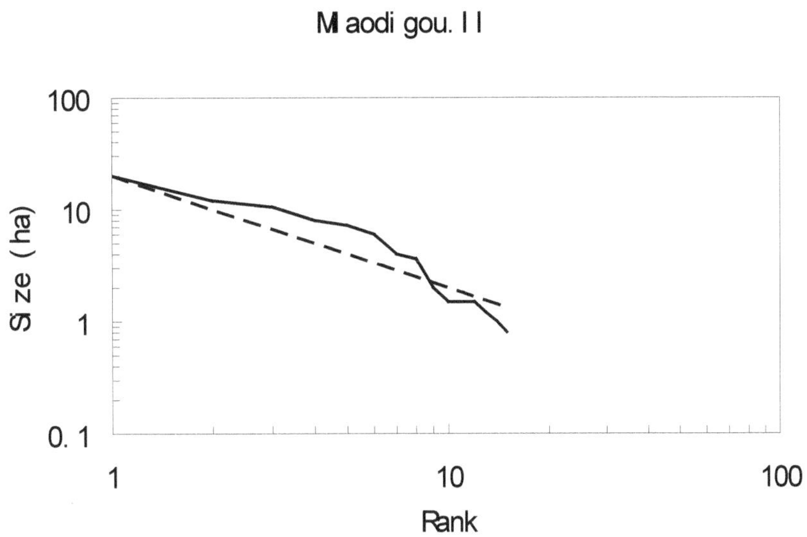

Figure 3-18. Rank-size distribution of Miaodigou II sites in the Yuanqu Basin.

two levels of settlement hierarchy akin to that of last period (Fig. 3-17). The rank-size curve is also similar to that of the Late Yangshao period, namely a little convex and near log-normal (Fig. 3-18). This indicates an integrated but weakly centralised polity, in which the central settlement, Dongguan may have still faced the competition from the other large settlement, Beibaotou (12 ha).

Five sites dating to Miaodigou II have been excavated in the Basin: Dongguan, Fengcun, Xiaozhao, Ningjiapo and Bancun. The data of the last two sites are not available at present. Fengcun (Shanxi Team 1987) and Xiaozhao (Shanxi Team 2001) only yielded several ash pits each, which provide little useful information for the analyses of communities. The following analyses will only focus on the Dongguan site (National Museum of Chinese History et al. 2001), which was excavated on a large scale and yielded comparatively abundant features and artefacts.

The excavation report of Dongguan divides the Miaodigou II remains of the site into three phases. Thirteen house foundations were uncovered for the three phases: six for the early phase, six for the middle phase, and one for the late phase. All of them are subterranean in structure. Except a few houses, most are less than 5 m^2 in size, and they are too small to function as household residences. Judging from their size and structure, most of them may have been storage or kitchen facilities. It is not clear about the residential pattern of this settlement.

Five pottery kilns (III Y1-Y5) dating to the early phase were found in Area III. These kilns are small, ranging from 0.5 to 1 m in diameter. Almost every kiln was linked to one large pit which functioned as the working area. Some storage facilities were distributed around the kilns. It is very likely that here was a specialised workplace for pottery production (see Chapter 6).

The ash pits, up to 161, accounted for the largest number among the features. Some of them, especially bell-shaped pits, may have functioned as storage facilities. These bell-shaped pits ranged from 1.5 to 4.7 m in diameter at the base, and the number of artefacts associated with the pits varied substantially, ranging from several to tens of ones. Given that these storage pits were linked with some households or kin-groups, those artefacts stored in bulk (or thrown out after feasting) were far beyond the everyday demand for any household, so they probably either served as the ritual goods for feasts, or served as trading goods for exchange. The difference of the number of artefacts stored in pits perhaps reflects differential successes in the social competition among individuals, households or kin-groups.

It is worth noting that some pits contained human skeletons, most of which were incomplete. Some of them were even mixed with animal remains (Fig. 3-19). It is obvious that the people buried in pits had humble status. Some of them were possibly sacrifices for ritual ceremony, or captives killed by the war victors. If it is the latter, violence and warfare may have become serious affairs in inter-group relationships.

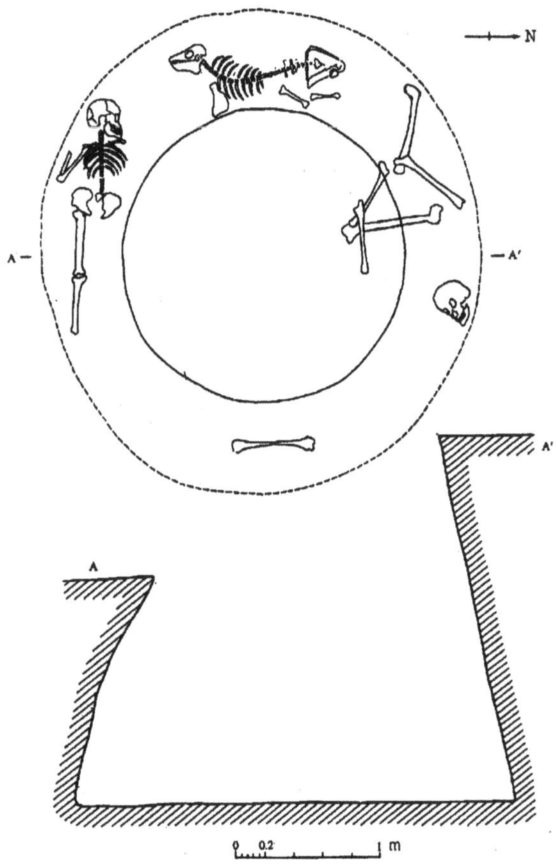

Figure 3-19. Plan and cross-section of the pit IIIH63 at Dongguan, late Miaodigou II period (after National Museum of Chinese History et al. 2001:fig.223).

Another noteworthy phenomenon is the appearance of "prestige goods" at Dongguan. A damaged elaborate stone item, *cong* tube, dating to the early phase (National Museum of Chinese History et al. 2001:223-224), is similar to jade *cong* tubes discovered in other contemporary sites, such as Taosi and Qingliangsi mentioned above. The jade *cong* tube is a typical kind of valuable prestige goods in some regions, especially in the Lower Yangzi River (the Liangzhu culture) (see Mou and Yun 1992). Though no jade objects were found at Dongguan, this stone *cong* appears to also imply the competitive emulation and aggrandising strategy linked with personal prestige and status.

In summary, the increase of the number of sites and the total occupation area, the enlargement of central place size, and the expansion of occupation space, all suggested an increase in population size and the amplification of the regional social scale. Nonetheless, all these changes were limited. Based on settlement patterns the social organisation in the Basin did not become more complex and integrated in this period than that of the previous period. The two-tiered settlement hierarchy probably still indicates a chiefdom-level polity and similar political hierarchy among villages as the previous period. On the other hand, based on material remains, conflicts or warfare may have become intensified in inter-group relationships, and competition for individual and/or group power and prestige may have led to the further development of social stratification during this period in the Basin. However, comparatively, the uneven development of social complexity probably still existed between some regions like Yuanqu and the other regions like Linfen (the Taosi case) and Ruicheng (the Qingliangsi case), where the highly complicated societies had developed (see the beginning of this section).

3.7. Longshan Period

During the Longshan period (ca. 2500-2000/1900 B.C.), several distinctive archaeological cultures existed in the Middle Yellow River valley. The Longshan complex in the Yuanqu Basin belonged to the Sanliqiao culture distributed in southern Shanxi and western Henan. The Sanliqiao culture was formed on the basis of the Miaodigou II culture and absorbed some cultural elements from several adjacent contemporary Longshan cultures (Dai 1998a).

During this period, the settlement and social patterns in the Central Plains experienced significant changes. A number of site clusters were distributed in circumscribed, semi-circumscribed and less circumscribed environments in various sizes, with two or three tiers of settlement hierarchy. The social systems that existed in these settlement clusters seem to involve both simple and complex chiefdoms (Liu 2004). A salient phenomenon for Longshan societies in the Central Plains is the unprecedented and striking evidence of warfare, indicated by the large quantities of weapons, a great number of abnormal dead humans buried in ash pits, and many walled sites (Yan 1997).

In the Yuanqu Basin, some new developing trends also occurred in the regional settlement patterns for this period:

(1) The total occupation area fell once again, and the

number of sites decreased slightly from 15 to 12 (Appendix 1), but the overall area of distribution did not shrink. These sites were dispersed widely over the landscape, and some sites were situated at the Western Tableland, Eastern Tableland, and margins of the Basin (Fig. 3-20).

(2) The central place shifted from Dongguan to Fengcun. Fengcun was located at the Western Tableland alone, so it was inconvenient for communication with other settlements. However, the geographic location and terrain of Fengcun was advantageous for its defence: it was situated at the high land and enclosed by the natural gulfs on three sides (see Fig. 3-20). Obviously, community security was the foremost consideration for the central place. This situation suggested that the inter-group conflicts or warfare had possibly become considerably greater than in early periods.

(3) The histogram indicates a three-tiered settlement hierarchy (Fig. 3-21). The largest site was Fengcun, measuring 30 ha in size. The second was Dongguan, up to 12 ha. Other sites were all less than 6 ha. The rank-size distribution shows a primate curve (Fig. 3-22), suggesting a highly integrated and centralised regional social system, in which Fengcun may have been the dominant centre of this regional polity.

In addition, it is worth noting that another contemporary large site, Longwang'ai, measuring 30 ha in size, appeared at the west of the Basin, and was about 15 km away from Fengcun (Shanxi Team 1986a). It is likely that a new regional polity had formed in the west of the Yuanqu basin for this period, and constituted the competition and threats to the polity in the Basin. Certainly, this proposition needs to be further tested through detailed surveys beyond the Basin in the future.

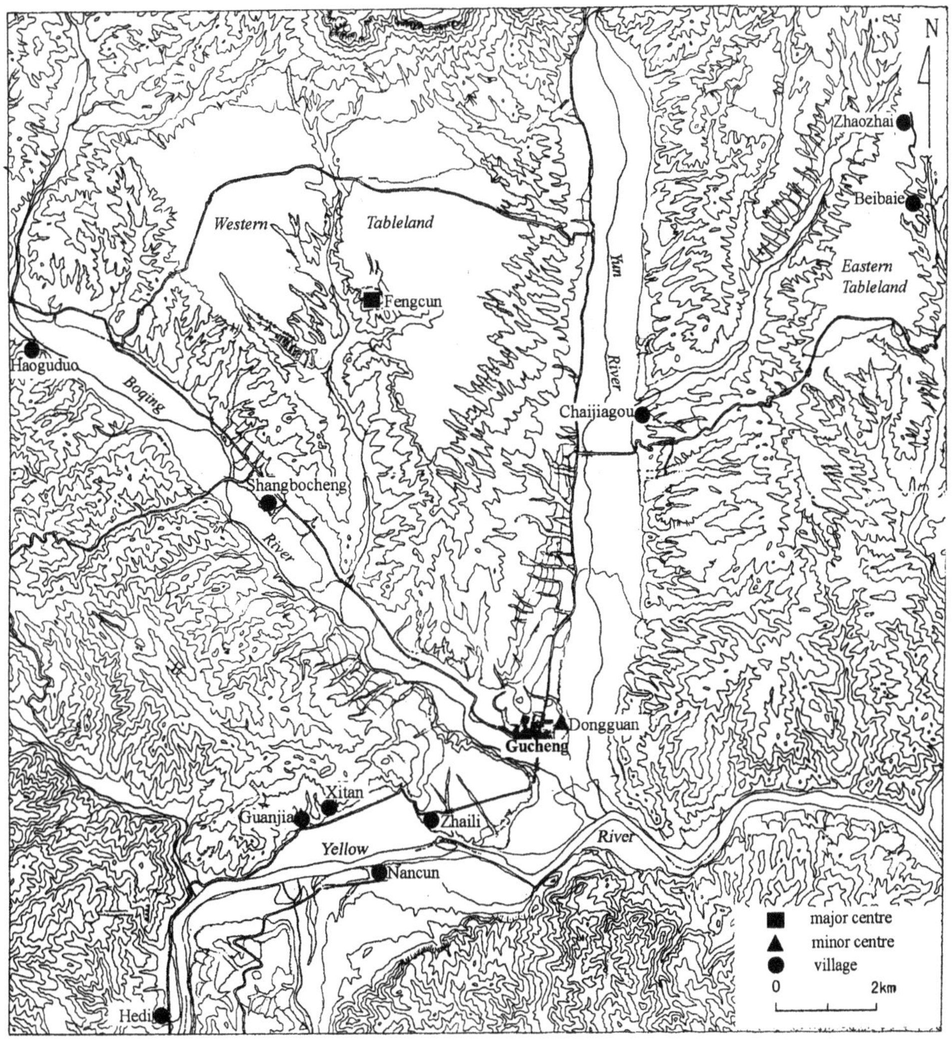

Figure 3-20. Distribution of Longshan sites in the Yuanqu Basin.

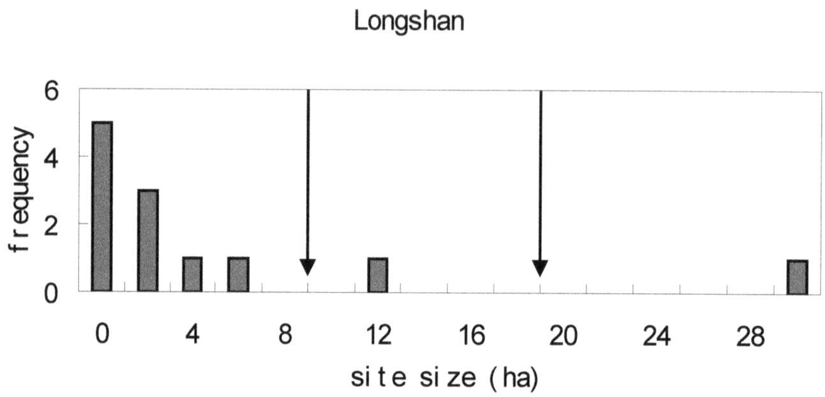

Figure 3-21. Histogram showing three levels of settlement hierarchy in the Yuanqu Basin, Longshan period.

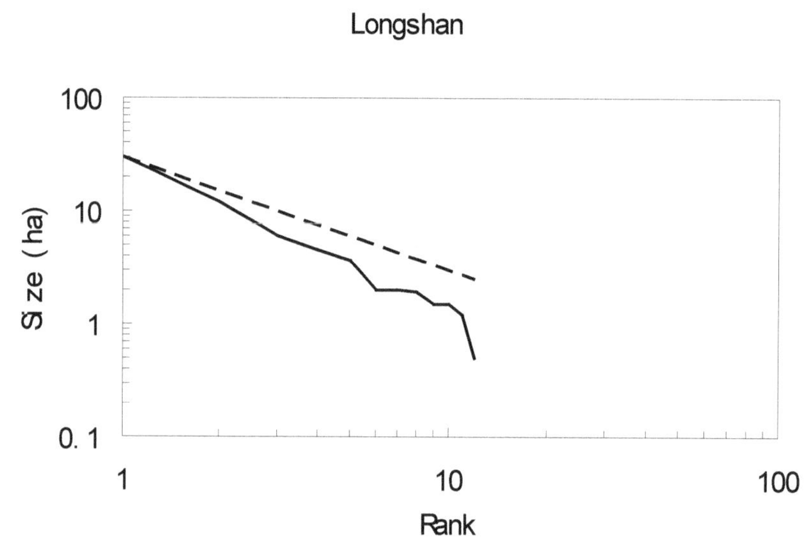

Figure 3-22. Rank-size distribution of Longshan sites in the Yuanqu Basin.

Of the Longshan sites in the Yuanqu Basin, Dongguan (National Museum of Chinese History et al. 2001) and Fengcun (Shanxi Team 1987) have been excavated. However, their settlement layout and structure are all unclear due to a small number of houses and burials unearthed. The major features uncovered in the two sites are once again ash pits, and some of them possibly still functioned as storage facilities, which provide little valuable information for the reconstruction of social structure and political economy. In addition, nine infant urn graves were dug out in Fengcun, which were arrayed in two rows closely. This mortuary tradition succeeding from the Yangshao culture probably reflected kinship structures that were still the basis of social organisations in this period. In Dongguan, a multiple-interment burial (IM5) including four people was uncovered. Two skeletons in the tomb were incomplete, lacking head and limb bones. There were no grave goods in the tomb (Fig.3-23). Besides, Dongguan yielded a large number of stone and bone arrowheads (64 in number). These are all probably evidence of violence or warfare.

In sum, the most salient changes of the social structure in this period were perhaps the increase of administrative hierarchy and the escalation of social integration. The population size and the scale of this regional polity appear to remain at the similar level as the previous period. There is not yet adequate evidence to indicate the further promotion of elite status and power, but intensification of conflicts or warfare was liable to result in more concentration of power, particularly when the society probably faced the threats and competition from another adjacent peer polity. It is possible that the complexity of the social organisation in the Yuanqu Basin developed farther in this period. This socio-political development in Yuanqu may have been consistent with most other regions in the Middle Yellow River valley, especially in the Central Plains (see Liu 2004:168-191).

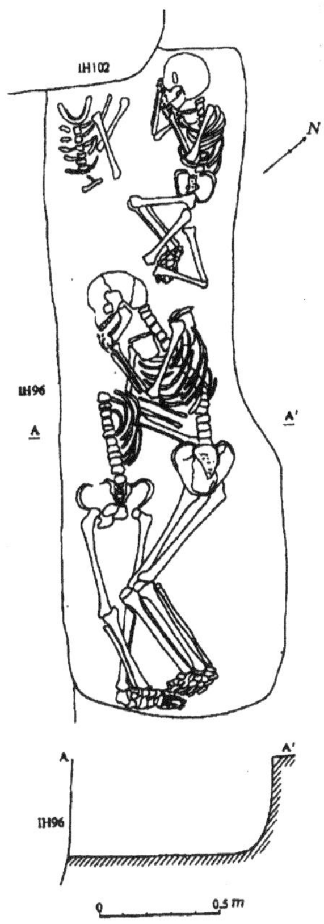

Figure 3-23. Plan of the tomb IM5 at Dongguan, Longshan period (after National Museum of Chinese History et al. 2001:fig.278).

3.8. Erlitou Period

The heartland of the Erlitou culture was in central Henan, where the Erlitou site was situated (near Yanshi, see Fig. 1-1). Beyond this region, there was a large periphery, including southern Shanxi and western Henan. The archaeological complex in southern Shanxi (including the Yuanqu Basin) did not directly succeed the local Sanliqiao culture of the Longshan period, but mainly came from the Erlitou culture rising in central Henan.

During the Erlitou period (ca. 2000/1900-1600/1500 B.C.), a state-level society emerged in the Central Plains. There is substantial controversy between Chinese and some Western archaeologists about the political and historical nature of the Erlitou culture (see Chapter One). As with most Chinese archaeologists, I believe that the Erlitou culture was a state-level polity, and the Erlitou site was its capital city (Institute of Archaeology 1999b).

Some marked changes had occurred in the regional settlement patterns in the Basin for this period:

(1) The number of sites increased considerably, from 12 to 20. Many of them were distributed densely, especially in the Yun and Boqing River valleys. Correspondingly, the total occupation area enlarged markedly (Appendix 1).

(2) Apart from the Eastern Tableland, these sites spread over all areas, including the margins of the Basin and the gorges of the Yellow River. However, there was only one site on the southern bank of the Yellow River (Fig. 3-24).

(3) The histogram shows a two-tiered settlement hierarchy (Fig. 3-25). The central place was still in Fengcun, but its area dropped to 20 ha. The Nanguan site was the largest (8.8 ha in area) among the rest villages. The rank-size curve is firstly near log-normal (slightly concave), and then a little convex, suggesting a well-integrated social system (Fig. 3-26).

Five sites containing the remains of the Erlitou period have been excavated in the Yuanqu Basin. Fengcun, Xiaozhao, Ningjiapo and Guanjia (Henan) provide little valuable information. Nanguan was excavated on a large scale for many years, and the excavated area was more than 1 ha. The excavation data from 1985 to 1986 have been published (National Museum of Chinese History et al. 1996). During my fieldwork in late 2003, I checked and looked through all unpublished excavation records from 1987 to 2003, especially focusing on those features, such as houses, burials, kilns and some pits. Unfortunately, the critical features of the Erlitou period in relation to settlement patterns are very few. Including the published data, there were only two small houses and one small burial discovered. The main features are still ash pits, which provide little useful information for the reconstruction of the settlement structure.

A significant phenomenon discovered at Erlitou component of Nanguan is that there were ditches extending at its northern and western boundaries (Fig. 3-27). Two parallel trenches (at the north, respectively with the width of four and eight meters) were linked with the other one (at the west, with the width of ten meters) at the north-western corner of the site. The Nanguan site was located at the eastern end of the upland, and it faced the bluffs at the east and south (see Fig. 3-24 & 27). The settlement was thus encircled by artificial and natural obstacles. The defensibility and security was evidently its special concern.

The Yuanqu Basin was close to the heartland of the Erlitou culture. As mentioned above, the archaeological assemblage for this period in the Basin did not develop directly from the previous Longshan culture—the Sanliqiao culture, but had many similarities with the Erlitou culture. This cultural replacement possibly implied the colonisation or direct domination from the core area of the Erlitou polity. The occupation of the Erlitou state in this region was probably related to the control and transport of copper and salt resources distributed in southern Shanxi (Liu and Chen 2000, 2003). In brief, the social organisation in the Yuanqu Basin may have been integrated into a state system during this period, and directly controlled and dominated by the Erlitou state as a regional administrative unit.

In this regional organisation, Fengcun was probably the administrative centre. Nanguan was contiguous to the Boqing and Yellow River with convenient transport, and had the defensive advantage with its special geographic

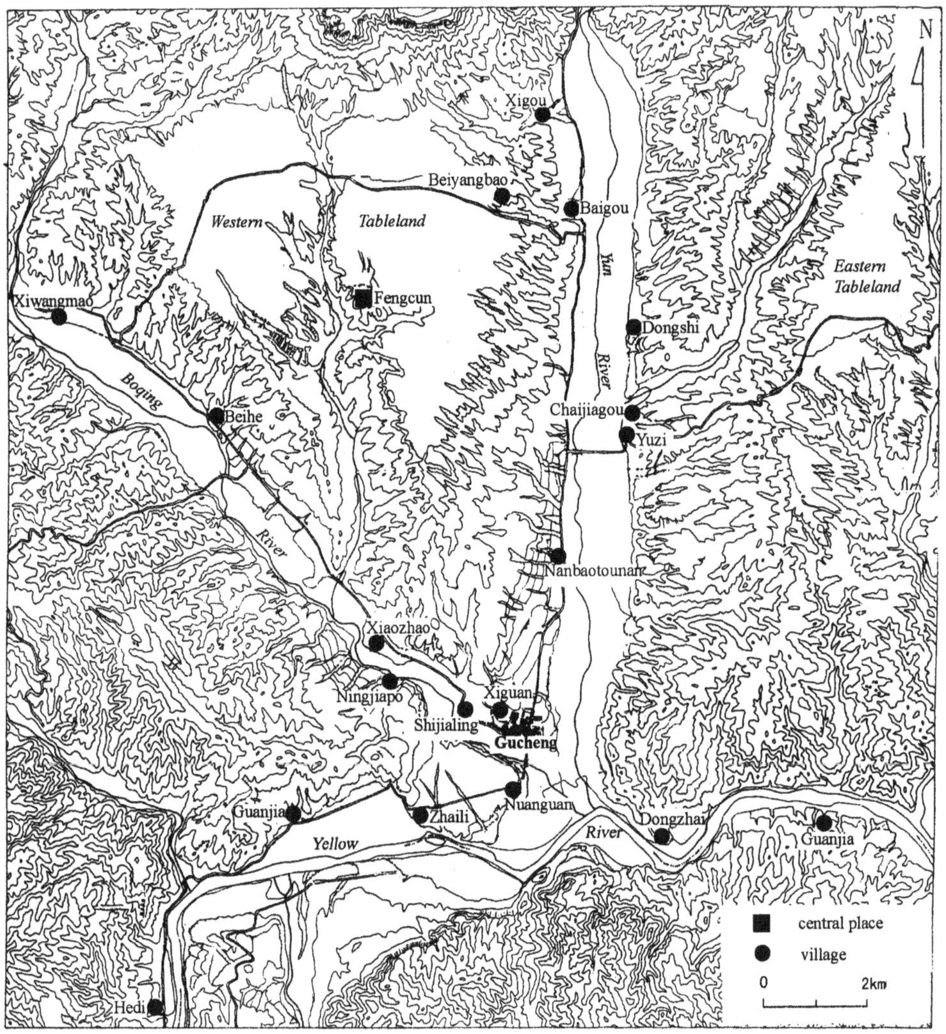

Figure 3-24. Distribution of Erlitou sites in the Yuanqu Basin.

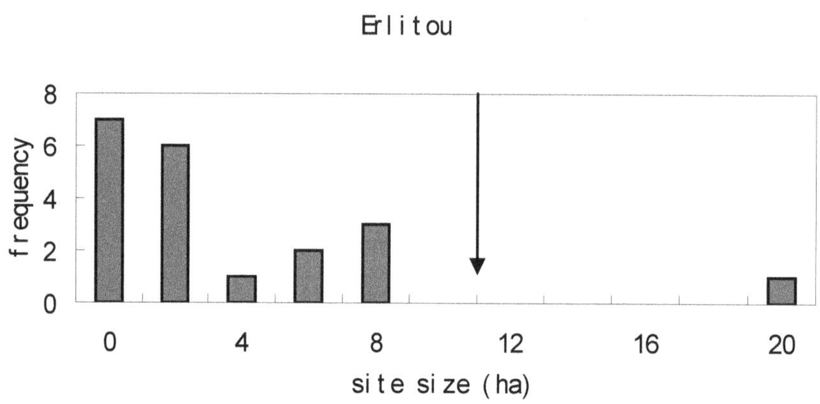

Figure 3-25. Histogram showing two levels of settlement hierarchy in the Yuanqu Basin, Erlitou period.

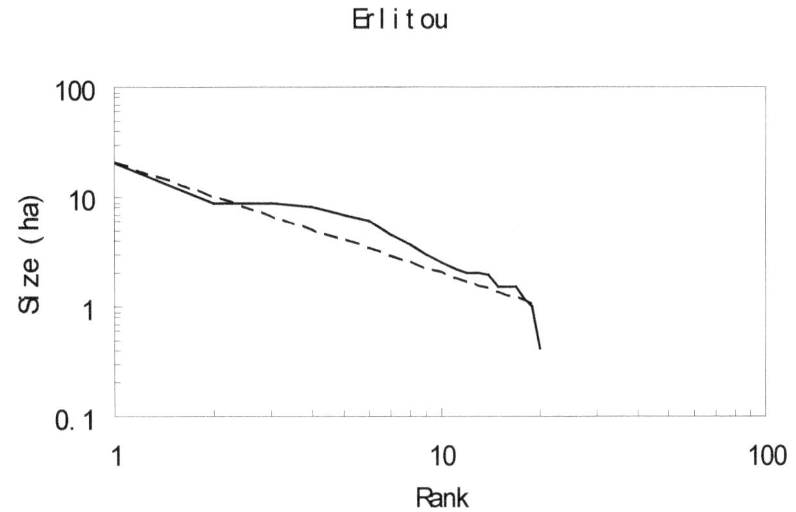

Figure 3-26. Rank-size distribution of Erlitou sites in the Yuanqu Basin.

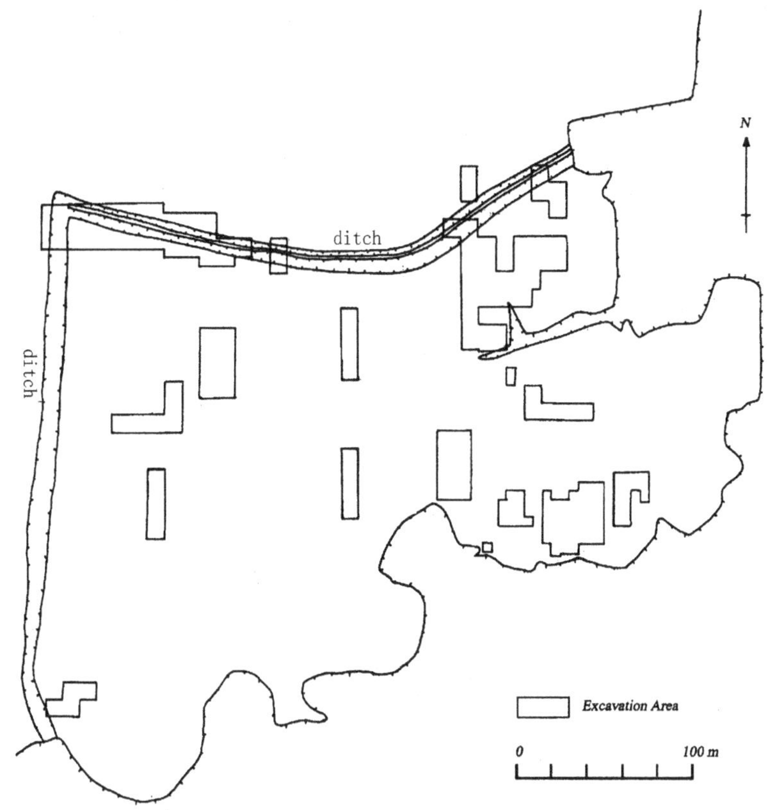

Figure 3-27. Outline of the Late Erlitou settlement at Nanguan (redrawn from the excavation record of Nanguan kept in the Gucheng archaeological workstation).

location and fortification, so it may have been in a critical position in controlling the transport of copper and salt through the river courses.

3.9. Erligang Period

During the Erligang period (ca. 1600-1300 B.C.), the Erligang culture replaced the Erlitou culture in the Central Plains. This cultural change coincided with the political events of the replacement of Xia by the Early Shang dynasty, as recorded in ancient texts (e.g., Sima 1982 [org. c. 100 B.C.]), so most Chinese archaeologists view the Erligang culture as the Early Shang dynasty culture. The heartland of the Erligang culture was still in central Henan Province, yet the first capital of the Erligang state was established in Zhengzhou (Henan Institute of Cultural Relics 2001; Zou 1980) or Yanshi

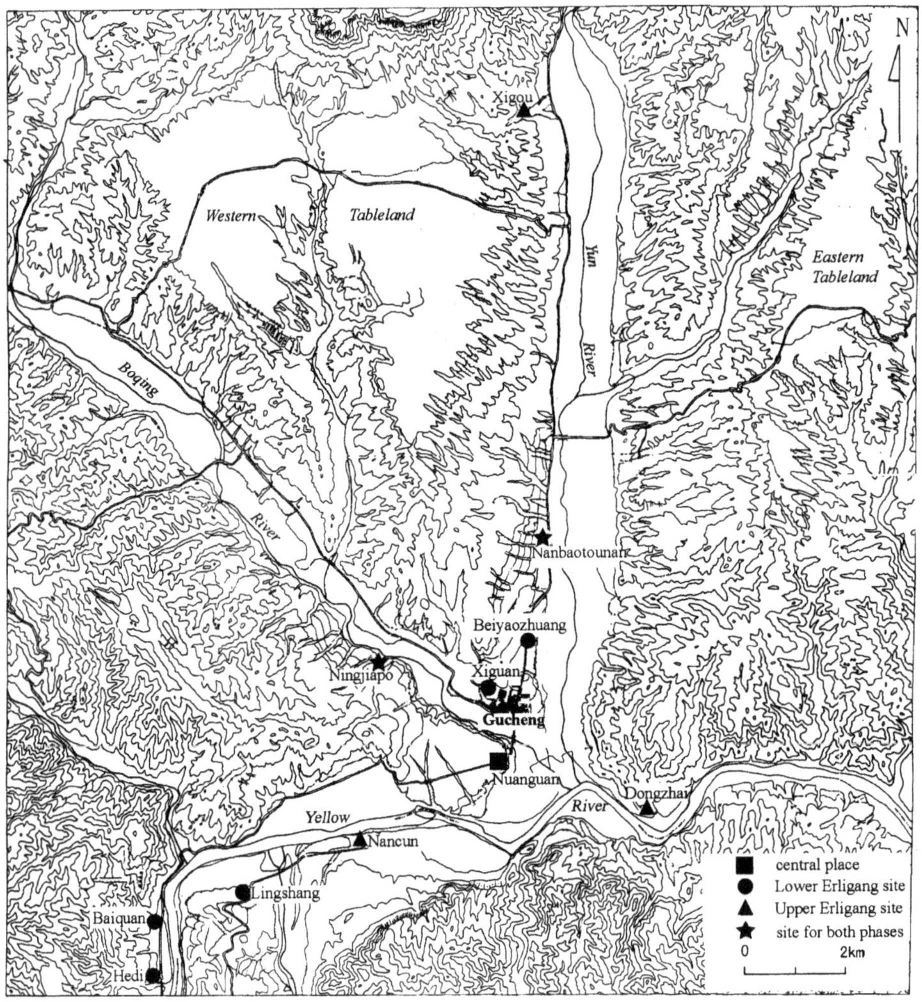

Figure 3-28. Distribution of Erligang sites in the Yuanqu Basin.

(Du et al. 1999) (see Fig. 1-1), and Erlitou was abandoned as a capital. Compared with the Erlitou culture, the Erligang culture had a much larger hinterland sphere beyond its heartland, still including southern Shanxi and western Henan.

This culture can be divided into two phases: the Lower Erligang and the Upper Erligang. The marked cultural attributes allow the specific chronological phase of every site dating to Erligang period in the Yuanqu Basin to be identified through excavated or surveyed data. Therefore, the following analyses will deal with the variations between two phases.

The Yuanqu Basin witnessed some critical changes of regional settlement patterns during this period:

(1) The number of sites and the total occupation area dramatically decreased. There was a total of 11 sites for the entire Erligang Period, eight sites for the Lower Erligang, and six sites for the Upper Erligang phase (three sites contains both phases) (Appendix 1).

(2) The distribution area of the sites in the Basin contracted. No sites have been found on the Eastern and Western Tableland. Most sites were distributed around the centre of the Basin and on both sides of the Yellow River (Fig. 3-28).

(3) The histogram indicates two levels of settlement hierarchy for both phases (Fig. 3-29). Although the Nanguan site was only 13.3 ha in area, it was still the largest settlement. Other sites were no more than 6 ha. The rank-size curves of both phases are very near log-normal, suggesting a highly integrated and centralised social system (Fig. 3-30), in which Nanguan may have become the central place in the Basin.

Of the sites excavated in the Basin, Nanguan and Ningjiapo contain the remains of the Erligang period. The settlement data of Ningjiapo is unavailable yet. Some results of the excavations at Nanguan have been reported (National Museum of Chinese History et al. 1996, 1997a, 1997b). In addition, I collected some useful data for settlement analysis from unpublished records of the Nanguan site during my fieldwork in late 2003.

After the Erlitou period, a critical change appeared in the Erligang settlement at Nanguan: a walled town was constructed in the Lower Erligang phase and used until the Upper Erligang phase (Fig. 3-31). The rammed-earth enclosure is slightly trapezoidal in shape, and measures 13.3 ha in area. The rammed-earth foundations of the north and west walls are still well preserved, while most of south and east walls

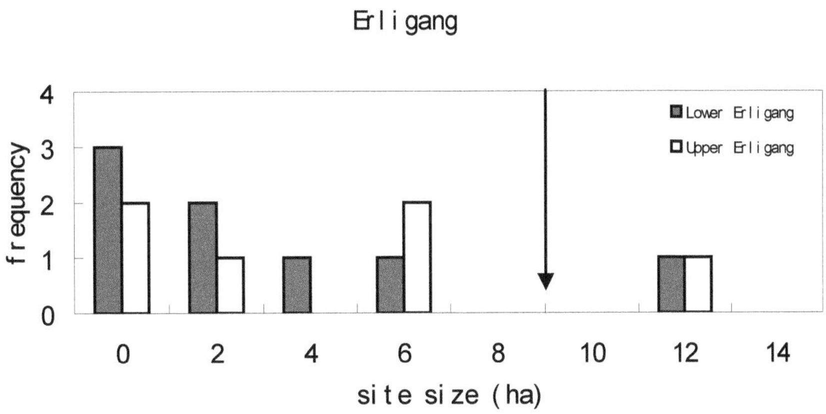

Figure 3-29. Histogram showing two levels of settlement hierarchy in the Yuanqu Basin, Erligang period.

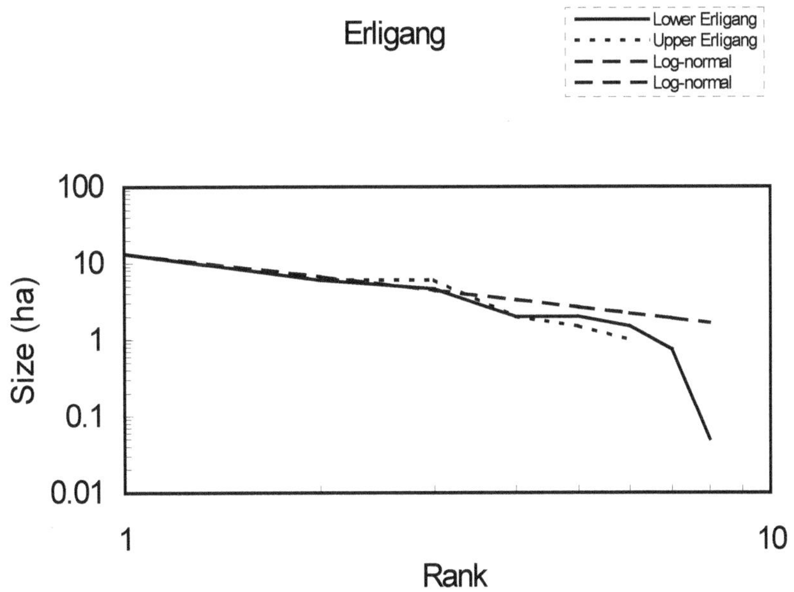

Figure 3-30. Rank-size distribution of Erligang sites in the Yuanqu Basin.

were close to the bluffs and destroyed by natural erosion and human damage. There were two parallel walls for the southern part and most of the western part of the enclosure. Beyond the west walls, a ditch paralleled them closely. This ditch was formed on the basis of the west moat of the Erlitou period by extending it northward, and the purpose was obviously to reinforce the defence on the west. During the Erligang period, two parallel north moats of the Erlitou period were filled with soil and functioned as a major road linking the west gate and the palatial district in the central section of the site. We found some residues of copper ore on the surface of the road, so it is very possible that this road was closely related to the transport of copper.

The whole walled town and the immediate areas outside the town were surveyed by our archaeological team with Louyang spades—a kind of coring device effective in surveying sites. There were hardly any features or cultural deposits of the Erligang period distributed outside the town and at the northern part of the town. Therefore, all excavations were concentrated in the southern part within the town (see Fig. 3-31).

A palatial district occupying an area of 0.45 ha was uncovered in the central portion of the walled town (see Fig. 3-31). It mainly consisted of two large rammed-earth platforms that were the foundations of the major buildings. The two platforms were encircled by a rectangle rammed-earth enclosure, which formed a segregated residential sector. This cluster of architecture may have been the residence and administrative centre of the elite who dominated the town, or even the whole Basin.

The general excavated area at the Nanguan site is over one ha. A large number of features of the Erligang period were distributed densely in the south-eastern section of the site, and the features in other areas of the site were comparatively sporadic. Nonetheless, it is very strange that only two houses of the Erligang period were discovered. It is possible that most remnants of residential buildings were either located outside

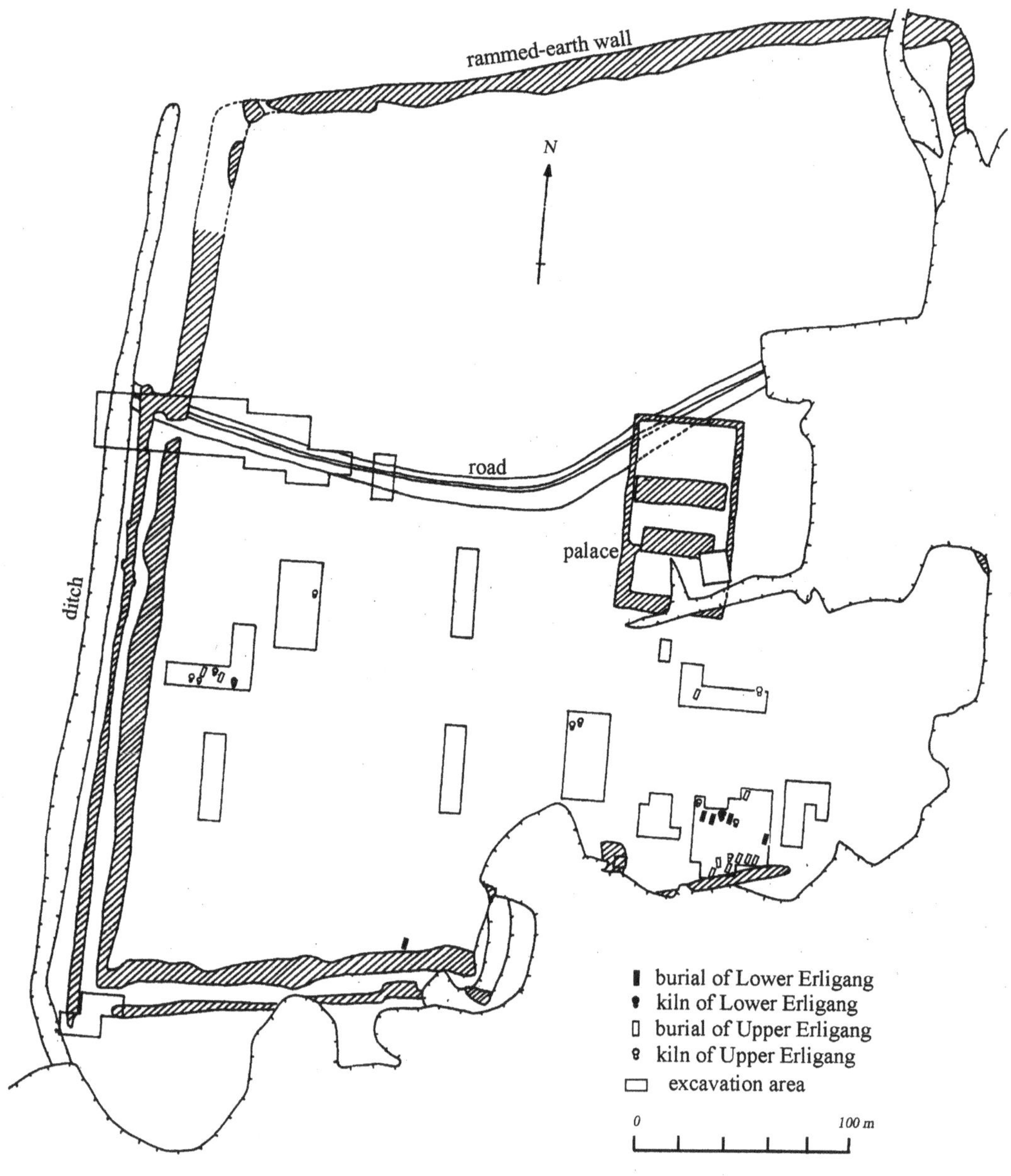

Figure 3-31. Map of the Erligang settlement at Nanguan (redrawn from the excavation record of Nanguan kept in the Gucheng archaeological workstation).

the excavated areas, or destroyed by later activities. The large quantities of ash pits, domestic refuse, and artefacts relevant to ordinary life and productive activities suggest that the south-eastern section of the site may have been a major residential zone for common people (see Chapter Seven).

Half a dozen of Lower Erligang burials and a dozen Upper Erligang burials were uncovered at Nanguan, most of them aggregated in the south-eastern part of the site (see Fig. 3-31). Differentiation of burial hierarchy is evident. A very small number of burials had elaborate grave goods. For example, M16, dating to the Lower Erligang phase, was a large tomb with a length of nearly three meters and a width of 1.67 m (Fig. 3-32). The central occupant of the tomb was an adult male, who was accompanied by a young female on the left. The major grave goods included two bronze vessels (*Jia* and *Jue* tripods) and a jade item, which were ritual items rather than utilitarian goods (National Museum of Chinese History et al. 1997a:19). The other elite burial, M1, dating to the Upper Erligang phase, yielded a dozen of grave goods including cowry, bone, pottery, jade, and bronze objects. The occupant of this tomb was an adult female (National Museum of Chinese History et al. 1996:211). Among other burials, several yielded a few pottery vessels, some of which were relatively rare and valuable wares (see Chapter Seven), so the occupants of these tombs perhaps had relatively high status in the society. For most small burials, there were few or no grave goods, and some skeletons were even incomplete. Accordingly, social stratification was manifested clearly in the burials.

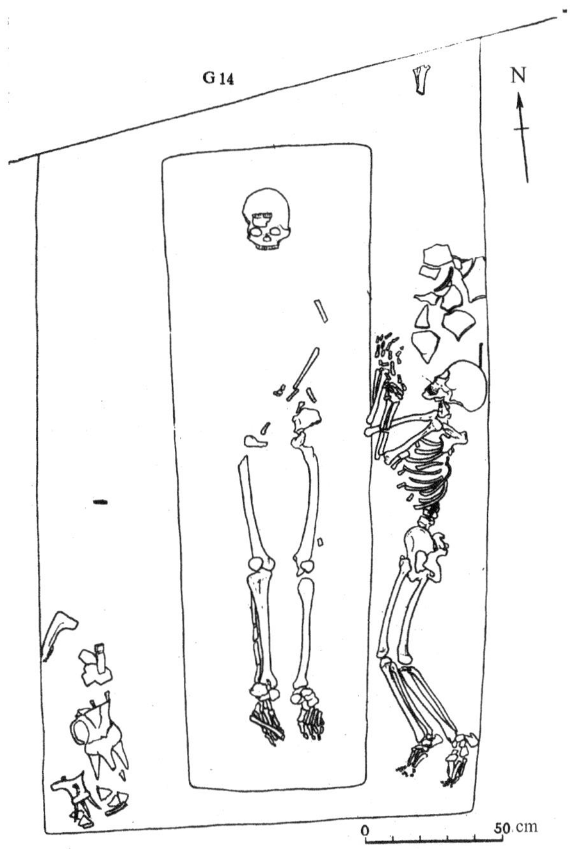

Figure 3-32. Plan of the burial M16 at Nanguan, Lower Erligang period (after National Museum of Chinese History et al. 1997a:fig.12).

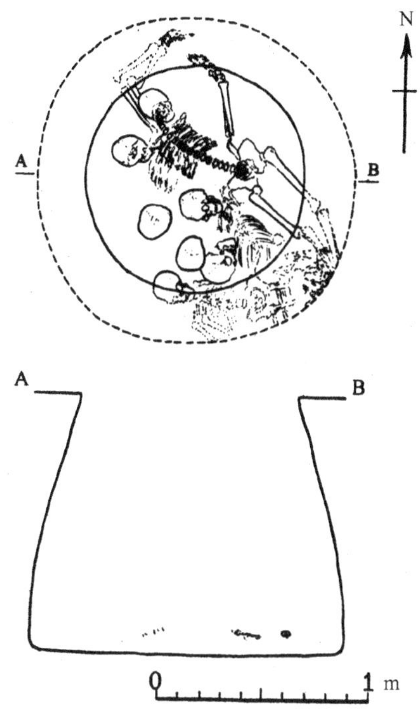

Figure 3-33. Plan and cross-section of the pit H353 at Nanguan, Lower Erligang period (after National Museum of Chinese History et al. 1997a:fig.13).

A great number of ash pits were found for both phases of the Erligang period. Some of them, especially the bell-shaped pits, were probably used as the storage facilities, many of which yielded abundant artefacts. Some bell-shaped pits were used for purposes other than storage. In a few such cases, each pit contained a single complete human skeleton (i.e., H248, H325, H392) (National Museum of Chinese History et al. 1996:212; 1997a:24).The people buried in the pits were probably members of the community with humble status. Two bell-shaped pits contained multiple incomplete skeletons. H353 (Fig. 3-33), which is 1.05 m in orifice diameter, 1.5 m in bottom diameter, and 1.2 m in depth, contained seven incomplete skeletons stacked in several layers. A bronze arrowhead was found to be inserted in the left leg bones of a skeleton (National Museum of Chinese History et al. 1997a:19). H661 (unpublished), which is 2.35-2.85 m in orifice diameter, 3.8 m in bottom diameter, and 3.25 m in depth, contained 14 skeletons, but there is only one skull among them. This pit yielded large quantities of pottery, stone and bone items, including many oracle bones. The people in the two pits were most likely captives or human sacrifices who were killed, buried, and associated with some kind of ritual activities.

A dozen pottery kilns were distributed in the south part of the site. It is curious that only one kiln was explicitly dated to the Lower Erligang phase, and all others belonged to the Upper Erligang phase. It is expected that some kilns beyond the excavated areas have not been discovered. Hence at least during the Upper Erligang phase, large-scale specialised pottery production may have occurred at Nanguan, and may have been manipulated by the elite of this dominant centre (see Chapter Seven).

The archaeological assemblage for this period in the Yuanqu Basin exhibited typical traits of the Erligang culture, but was not directly descendent from the Erlitou culture. Therefore, similar to the previous period, this cultural replacement may have been the material reflection of regime change. In other words, the Erligang state took over the regional organisation in the Yuanqu Basin from the Erlitou state and directly controlled it. This suggestion can be supported by some discoveries in the Nuanguan site. The ritual bronze vessels unearthed in two elite tombs (see discussion above) were very similar to their counterparts discovered in the core area of the Erligang state in central Henan (e.g., Henan Institute of Cultural Relics 2001). As there is no evidence of casting ritual bronze items in Nanguan, these ritual goods were probably obtained directly from the royal court.

In general, although the sharp decrease in the number of sites, the total occupation area, and the central place size indicate a substantial decline of the population and social scale during the Erligang period in the Yuanqu Basin, the regional polity in the Basin was still well integrated and incorporated into a greater state social system.

According to ancient Chinese literature, regime change between the early states was accompanied by large-scale wars (e.g., Sima 1982 [org. c. 100 B.C.]). It is likely that the occupation of the Erligang state in the Yuanqu Basin was carried out through military rather than peaceful means. This point can be confirmed by the evidence of war and violence discovered in some pits containing a quantity of human remains (see above). This was perhaps one of direct and crucial reasons leading to the abrupt decline of the population and scale of regional organisation for this period.

During this period, the Nanguan site may have become the regional administrative centre. It was still closely associated with controlling and procuring copper and transporting salt (Liu and Chen 2000; 2003:105), and its defensive and military function was further strengthened by the walled enclosure. In this control centre, the magnificent palatial buildings displayed unprecedented residential segregation between elite and common people, and social stratification was also adequately expressed in mortuary practice. Besides, the Nanguan walled town may have been not only the political and military centre in the Yuanqu Basin, but also a significant base of craft production.

The Shang culture suddenly declined during the Late Shang period in southern Shanxi, and there were few sites of this period discovered in this region. In the Yuanqu Basin, there is only sparse evidence of Late Shang remains (National Museum of Chinese History et al. 2001:429). One of explanations for this archaeological phenomenon is that the major sources of copper and salt had shifted from southern Shanxi to other places during the Late Shang period, so the Late Shang dynasty gave up the control to southern Shanxi, including the Yuanqu Basin (Liu and Chen 2000, 2002, 2003).

3.10. Summary and Discussion

The analyses above exhibit a long-term evolutionary process from simple to complex societies in the Yuanqu Basin. I try to reconstruct the socio-political forms for each period through settlement pattern studies. However, we must realise that the settlement data are often inadequate, especially on the community level, so some analytical results are inferential rather than conclusive, and some of them are implicit rather than explicit. This nature of the data leaves a lot of issues that need to be tested, confirmed and resolved for the future studies.

The settlement hierarchy analysis and the rank-size analysis play an important role in reconstructing socio-political structure of the regional settlement system in this study. However, it needs to keep in mind that these two methods should not be used in isolation, and they can make sense only when they are used in combination with other archaeological data and analytical means. For example, dependence on settlement-political hierarchical models leads to a conclusion that the social complexity during the Erlitou and Erligang periods declined in the Yuanqu Basin because there were only two levels of settlement hierarchy, but the social systems for the two periods became actually more complicated since they were integrated into the greater state societies. Nonetheless, sometimes I have to infer the social form or nature mainly relying on these two methods due to very limited archaeological data, such as for the settlement analysis of the Late Yangshao period in the Basin. This is a problem that needs to be particularly addressed in the future studies. On the other hand, almost all archaeological case studies confirm that the emergence of settlement hierarchy (at least two-tiered settlement hierarchy) in a specific and clearly defined region indicates a complex society rather than a simple society (see Chapter Two). Therefore, I believe that some inferences about the basic social form (simple or complex society) mainly based on settlement hierarchy should be viable in this study (such as the inference of the social form of Late Yangshao).

In summary:

The social structure of the Early Neolithic in the Yuanqu Basin is unclear due to the lack of archaeological sites for this period. From the Peiligang to the Early Yangshao period (ca. 6200-4100 B.C.), the settlements in the Basin were few in number and small in size, and there was no hierarchical differentiation within and among communities. Each village may have been autonomous, and organised on the basis of kinship. These independent social organisations may be referred to as "local groups", which approximate the "tribal societies" of Service's evolutionary model (Service 1962), since "tribes" cover most "local groups" (Johnson and Earle 2000:129). On the other hand, the social structures possibly were not static during this long period, and some changes probably had occurred, but we cannot yet specify them due to the limitation of data.

The Middle Yangshao period (ca. 4100-3500 B.C.) may have witnessed the emergence of the first complex society in the Yuanqu Basin. The large-size central place and three-tiered settlement hierarchy indicate a well-integrated and stratified social system, and the supra-village regional polity may have formed. Nonetheless, there is little evidence signalling obvious high-ranking individual status, so this sort of regional polity may be characterised as a "group-oriented chiefdom". This suggestion can be supported by the large central settlement beyond the Yuanqu Basin (i.e., Xipo in Lingbao), in which the most salient features were public buildings rather than valuable personal possessions.

From the Late Yangshao to the Miaodigou II period (ca. 3500-2500 B.C.), the social patterns in the Basin seem to have many similarities. Compared with the Middle Yangshao period, the population size and the social scale substantially declined (though there was a upward fluctuation between them), and social integration was also relatively weak for these two periods in the Basin. Social complexity may have experienced uneven development among different regions in the Middle Yellow River valley. On the other hand, the social organisation of the two periods in Yuanqu still displayed some characteristics of chiefdom-level societies, and inter-group conflicts and competition for individual/group power and prestige may have become increasingly marked over time.

Although the social scale in the Yuanqu Basin may have not amplified during the Longshan Period (ca. 2500-2000/1900 B.C.), the regional polity probably became highly integrated once again. The conflicts and warfare possibly existing between two adjacent polities may have intensified the military competition for political power among elites. This condition could give rise to the further development of social complexity.

From the Erlitou to Erligang period (ca. 2000/1900-1300 B.C.), namely the Early Bronze Age, social organisation in the Yuanqu Basin may have been integrated into two successive state systems. The residential segregation and mortuary practice in the Erligang period at Nanguan also strongly suggest a highly hierarchical and stratified social system. The expansion of the two early states to the Basin was possibly directly related to the procurement and transport of copper and salt resources available in southern Shanxi. This region therefore probably played an important role in forming the highly integrated state systems.

The social changes in the Yuanqu Basin exhibited an evolutionary process from tribe, chiefdom to state societies, or from local groups to regional polities (chiefdoms and states), so it conforms to the evolutionary models of social development (Johnson and Earle 2000). On the other hand, we should also be aware of the complexity of the evolutionary process. For example, the first complex society probably emerging in the Middle Yangshao period involved a three-tired settlement hierarchy (two-level control hierarchy) and was well-integrated, but after that, during a considerably long time (from the Late Yangshao to Miaodigou II period), the regional polities in the Basin possibly remained at a weakly integrated level and had only two tiers of settlement hierarchy (one level of decision-making hierarchy). Not until the Longshan period did the social organisation in the Basin begin to become more complex.

An intriguing phenomenon is that the number of sites and the total occupation area displayed regular periodic changes from the Early Yangshao to the Erligang period in the Yuanqu Basin (see Appendix 1), and correspondingly the population size and social scale may have experienced the same developmental cycles. However, at present we cannot yet identify the principal factor(s) leading to these changes.

A notable phenomenon is that the emergence of complex society in the Yuanqu Basin was closely linked with population growth. The formation of the first regional polities (chiefdoms) during the Middle Yangshao period in the Central Plains accompanied a sharp increase of the number of sites, the total occupation area, and relevant population sizes (Dai 2004), including the Yuanqu Basin and Zhudingyuan (Ma 2003) addressed in this chapter. The relationship between demographic factors and social complexity has been an exceedingly controversial issue. Many take opposing views as to whether population growth is the reason for the result of social complexity (e.g., Blanton et al. 1993 ; Carneiro 1970; Feinman 1995; Flannery 1972; Johnson and Earle 2000; Sanders 1984). Likewise, it is not clear whether or not population pressure directly resulted in the origin of complex societies in the Central Plains, including the Yuanqu Basin, although population growth seems to closely correlate to the development of social complexity. In addition, at present we have no sufficient evidence to elucidate why the population dramatically increased during the Middle Yangshao period.

In general, on the basis of current data in the Yuanqu Basin, it is impossible to resolve dynamics of the emergence and development of social complexity, which still remain to be explored in the future studies.

Chapter Four

Modelling Pottery Production: Theories, Methods, and Data

4.1. Introduction

As stated in Chapter One, the main aims of the pottery studies in this thesis are to reveal the changes of modes of production from the Neolithic to the Early Bronze Age in the Yuanqu Basin, to evaluate the development of specialised production over time, and to examine interrelationships between the organisation of pottery production and the degree of social complexity. In this chapter, I discuss some theoretical and methodological models in relation to pottery production, especially the mode of production and specialisation, and then detail the approaches to pottery data collection and analyses.

Craft production is one important component of socio-political economic systems. Cathy Costin suggests that the production system consists of six constituent components: *artisans, means of production, organisation and social relationships of production, objects, relationships of distribution,* and *consumers* (Costin 2001:277). These components are doubtless interconnected. It is impossible to study any one aspect completely independently of other components of production. But at the same time, every component can constitute a relatively independent study theme in the context of the whole production system. This dissertation focuses on the organisation and social relationships of pottery production, and here, I mainly address theories and analytical methods related to this aspect.

4.2. Mode of production, Specialisation, and Social Complexity

4.2.1. The Mode of Production and Socio-political Structure

In recent years, an influential model exploring the relationships between pottery production and social complexity stems from a paper by Rice (1981). In her evolutionary model, increasing social complexity is accompanied by the concurrent establishment and development of specialised production, which is embodied in an increase in varieties of ceramic wares (such as shape and/or decorative classes), especially in prestige vessels, and an increase in standardisation of paste, size and morphological attributes within shape classes, particularly in utilitarian wares (Rice 1981:222-224).

Rice's model indicates a crucial starting point for evaluating relationships between modes of pottery production and socio-political forms and provides a valuable approach to examine the change of pottery production. She develops some essential terms such as specialisation, standardisation, diversity and variability, and discusses the interrelations among them (Rice 1981, 1984, 1987, 1989, 1991). These concepts will play critical roles in the following ceramic studies. I will test the validity of her model through investigating the change of pottery production in the Yuanqu Basin.

The mode of production can be manifested by productivity and the relationship of production. As Rice (1987:182) notes:

> Study of the mode of pottery production is based on interrelated determinations of how the pottery is made, who makes it, and for whom it is made. Pertinent questions thus relate to manufacturing technology, the role and status of producers, the integration between tasks, the organization of producing units and their relation to the overall economic organization, and the relation between producers and consuming groups.

The organisation of production, as one critical component of the mode of production, has been a focus of pottery study for archaeologists and ethnologists. Two popular schemes of organisation of pottery production are proposed by van der Leeuw (1977) and Peacock (1982), drawing on the ethnographic record. Rice (1987:184) suggests four modes of production in prehistoric contexts based on their schemes: *household production, household industry, individual workshop industry, and nucleated workshops.* These four modes of production are actually the first four types in Peacock's scheme. These concepts are also further developed by other scholars, such as Pool and Santley (1992:213-214). These models may be simply summarised as follows.

Household production. This is the simplest mode of production, in which each household makes pottery occasionally for its own use. Technology of production tends to be simple, and there is a lack of investment in permanent facilities, such as wheels and kilns. Therefore it is exceedingly difficult to identify this production mode in the archaeological record.

Household industry. In this mode pottery-making is a part-time activity conducted by a few relatively skilled artisans within a domestic context. Pottery is produced for profit as a secondary but possibly important source of income to supplement the household economy. Some specialisation for specific ceramic categories may happen. Investment in equipment is perhaps slight, but meantime more complex implements and facilities such as kilns may be expected.

Individual workshop industry (*individual workshops*). "The division between this mode and the previous one is very fine" (Peacock 1982:9). However, in this case pottery is usually produced in isolated and dispersed workshops for exchange as a vital or even primary source of subsistence. Potting may be still part-time or seasonal activity, but the significant capital investment in permanent facilities (kilns, wheels, etc) is more visible.

Nucleated workshops. "In this mode individual workshops are grouped together to form a more or less tightly clustered industrial complex" (Peacock 1982:9), in which pottery production is a major economic activity and the cooperation in resource procurement or product distribution may exist. Production is usually year-round (or seasonal for the impact of climate) and products are highly standardised and of high quality. Competition results in extensive technological investment in pottery manufacture.

The last four modes of production in Peacock's model are the *manufactory*, *factory*, *estate production*, and *military and other official production* (Peacock 1982:9-11). They relate much more to historical contexts and have little to do with my study, and need not be considered further here. Among other models relevant to the mode of production, Costin (1991:8-9) also presents an elaborate eight-part typology for the organisation of specialist production, but this typology is too detailed to pertain the archaeological analyses in this study.

For whatever typology of production modes, it is not easy for archaeologists to link the specific organisation of production with the particular social form for the following reasons:

(1) It is very difficult to explicitly identify the organisation of pottery production in archaeology. As Rice states, "If production were treated alone, strictly independent of distribution, one could see that ......there are basically two kinds of production arrangements, household and workshop". Household production, household industry, and to some extent individual workshop industry "are likely to leave little trace archaeologically because of lack of material technological investment, low intensity of activity, and isolated or dispersed geographical distribution." For individual workshop industry and nucleated workshops, "the extent to which these latter could be identified in prehistory is questionable" (Rice 1987:186-187).

(2) The development of the mode (organisation) of production may not conform to a uni-linear evolutionary model, although the change from household to workshop production might represent the general tendency. There is no simple correspondence between organisational typologies of production and social patterns. In any one society, there may exist different modes of production (see Costin 2001:274; Sinopoli 1991:102). For example, the Middle Classic Period at Matacapan, south-eastern Mexico witnessed diverse modes of production from household industry to workshop industry and even nucleated industry (Santley 1994).

(3) The complexity of organisation of production cannot be covered completely by these typological schemes. "It is somewhat difficult to place various community-based specialisations into this scheme" (Rice 1987:187). There may be a great deal of variability within and/or beyond these foregoing schemes.

Despite these difficulties, archaeologists can still make contributions to the study of modes of pottery production, by focusing on those problems that can be solved with archaeological evidence in the specific socio-economic context. One advantage of archaeology is that it can provide diachronic long-term insights into the change of some domains of human societies. Assessing the development of specialisation has been one of the major preoccupations of studies of the modes of production over the past two decades.

4.2.2. Specialisation

Rice suggests three categories of specialisation: *site specialisation, resource specialisation, and producer specialisation* (Rice 1987:190-191; 1989:110; 1991:262-263). Most people refer to craft specialisation as producer specialisation. There are a number of definitions of specialisation among researchers (e.g., Blackman et al. 1993:60; Costin 1991:3-4; 2001:275-276; Rice 1991:261; Stark 1992:184; Stein 1996:25; Stein and Blackman 1993:29). I adopt the one suggested by Costin (1991:4):

> I would argue that specialisation is a differentiated, regularized, permanent, and perhaps institutionalized production system in which producers depend on extra-household exchange relationships at least in part for their livelihood, and consumers depend on them for acquisition of goods they do not produce themselves.

In essence, craft specialisation refers to production for consumers beyond households of producers (Clark 1995; Clark and Parry 1990; Costin 1991; Rice 1987; Underhill 2002). Rice (1991:259) argues that the origins of specialised pottery production address three kinds of causal factors: "sociopolitical factors (the development of elites and centralized decision making ...); economic factors, especially agricultural intensification and competition ...and the general relationship between sedentism and productive intensification ...; and ecological/environmental factors." This interpretation is similar to Costin's proposal (Costin 2001:307; Rice 1991:260). Above all, both Costin and Rice and many other scholars (for example, Feinman et al. 1984; Hagstrum 1985; Sinopoli 1991) emphasise the significant role of population growth and nucleation, which give rise to agricultural intensification, the extension of scale of craft production, and the increase of productive efficiency, and ultimately result in specialised production.

For the appearance of attached specialists, socio-political factors play a critical role. By sponsoring and monopolising the production of special goods symbolizing power and wealth, the elite can enhance and legitimate their social status and authority (Brumfiel and Earle 1987; Costin 1991).

Similar to Rice (1981), many scholars view craft specialisation as a concomitant of social complexity, and postulate that there is a positive correlation between craft specialisation and social complexity (i.e., Earle 1987; Stein 1996). However, Rice's model is also opposed by some scholars who deny the close correlation between social complexity, social stratification, and craft specialisation. "Cross-cultural study shows that craft specialisation may appear in what the author's evolutionary scheme would call a 'simple' society" (Davis 1981:228). Hodder stresses the role of the particular cultural context, and argues that economic specialisation is not a general concomitant of social complexity (Hodder 1981:231). Some other scholars also contend that craft specialisation is not linked only to complex societies; it is also likely to be found in non-stratified societies (Clark and Parry 1990; Cross 1993; Underhill 2002:4). I also argue that although there is a certain correlation between craft specialisation and social complexity, it is incorrect to simply say that craft specialisation is a concomitant of social complexity (see the following studies).

The relationship between specialisation and division of labour is a significant subject that has not been studied sufficiently. The supra-household division of labour is an essential component of specialist production. The transformation from intra-household divisions of labour to extra-household divisions of labour is likely to take place in "middle-range" societies (Costin 2001:309). If so, it is possible that the initial specialisation appears in a non-stratified society. This proposition deserves further testing in this study.

In addition, I argue that a pivotal point is that we should understand the exact implication of specialisation, when we use this term to specify some kind of mode of production, and when we discuss the origin and development of specialisation. Costin has pointed out clearly, "Specialization is not a single organisational state, ...it has degrees", and "specialization can be organized in many ways; there are many types of specialization" (Costin 1991:4). For example, we can distinguish between "household specialisation and workshop specialisation", "part-time specialisation and full-time specialisation", or "independent specialisation and attached specialisation". Accordingly, it is necessary and essential for archaeologists to identify distinct degrees and/or types of specialisation when they try to work out interrelation between specialisation and social complexity.

4.2.3. A model for Assessing Specialisation or the Organisation of Production

Costin (1991) establishes four parameters that identify and explain the organisation of production: the *context* of production (independent or attached specialisation), the relative regional *concentration* of production (from dispersed to nucleated production), the *scale* of the production units (from small, kin-based units to factories), and the *intensity* of production (part-time or full-time specialisation). I will briefly discuss these four parameters mainly in the light of Costin's model (Costin 1991; also see Costin and Hagstrum 1995), and then focus on the context of production which has received extensive attention and discussion in recent years.

The *context* of production reflects the nature of the demand for a particular good. Independent producers tend to manufacture utilitarian goods (or subsistence goods) for a general market and serve in household maintenance. Attached specialists, who are sponsored and managed by elite or governmental institutions or patrons, usually produce prestige goods (or wealth goods) for a limited portion of the population to maintain political power and enforce social distinctions. The development of independent production is usually affected by market demands, population density, resource distribution, and so on. By contrast, attached specialisation is primarily influenced by social and political factors. It evolves along with social inequality, as a means for elites and governments to control the ideology and technology of power by monopolising special, high-value goods.

The *concentration* of production describes the spatial relationship between producers and consumers. At one extreme, artisans are evenly distributed throughout the population. At the other extreme, specialists are aggregated at a single community within a region, and consumers get products through regional or interregional exchange. The spatial arrangement of production activities is affected by a number of natural and social factors. It is firstly associated with the context of production. For independent production, the degree of concentration of production relies on environmental diversity, territoriality, transport costs, market setting, and relationship of cooperation among producers. For attached specialisation, the relative nucleation of production depends on overall control over raw materials, technology, production processes and final distribution, so most activities of attached production are conducted in nucleated settings that can facilitate monitoring and management by elites or patrons. My study will demonstrate that the spatial pattern of pottery production in the Yuanqu Basin is more susceptible to socio-political changes rather than other elements, for example, the clay resource is always accessible physically in the loess area of the Middle Yellow River, and is not an restricted element for the distribution of pottery production.

The *scale* of production refers to the group size and social relations of producers. At one extreme is household production within a domestic context. At the opposite extreme is the workshop or factory in a non-industrial background. There is no necessary link between the scale and context of production. Both independent and attached production could be large and small. For independent production, efficiency is the primary factor determining the scale of production. If a larger workshop enables producers to exploit expensive technology and marketing strategies more effectively, and get higher output, the scale of production is apt to rise. For attached specialisation, the scale of production depends primarily upon the need for supervision and required output level. In addition, I argue that population density and social demands are also essential factors to affect the scale of independent production. In practice, a variety of organisations of production can reflect distinct scales of production to substantial extent.

The *intensity* of production reflects the relative amount of time producers spend on craft production. At one extreme is part-time production where craft production is used to meet basic domestic demands or supplement domestic income. At the opposite extreme is full-time specialisation where specialists focus on a single craft or task as the primary source of income. Several economic factors, such as efficiency, risk, and scheduling, determine the intensity of independent production. The competitive pressure in these aspects can compel part-time artisans to turn to full-time production, by which they can increase productive efficiency, minimize risk, focus on their craft task, and finally get the competitive advantage. In contrast, attached specialists tend to be full-time workers, because they are sponsored and controlled by elites or patrons, and can be engaged in the production of high-skill wealth items more efficiently.

Generally, the intensity of production is a productive parameter for measuring the degree of specialisation. However, it is necessary to clarify the implication of "part-time" and "full-time" in case they are misunderstood. Part-time craft production includes intermittent, occasional, and seasonal activities, which are usually performed in non-agricultural time, and are usually secondary in the subsistence economy. In contrast, full-time production is usually year-round work and in the primary position in subsistence economy. Nevertheless, "true full-time specialised craft production is rare in the preindustrial world" (Costin 2001:280). It is possible that full-time specialists would have to stop craft production temporarily for specific natural, social or economic factors. For example, full-time potters are likely to stop their work during the busiest agricultural season or the rainiest season, especially for independent specialists. "Full-time" is used within this sense in this study.

These four parameters can make the assessment of productive modes more reasonable by reflecting the complexity and flexibility of the mode of production.

Finally, I intend to discuss a few controversial issues linked to the context of production, a crucial concept depicting the mode of production.

Many scholars make a dichotomous differentiation between independent and attached production (Arnold and Munns 1994; Brumfiel and Earle 1987; Costin 1991; Costin and Hagstrum 1995; Lewis 1996; Rice 1991; Stein and Blackman 1993, etc.). However, this clear distinction is increasingly questioned in recent studies, and some scholars contend that the productive context is more varied and complex than this dichotomy (Costin 2001; Lass 1998; Spielmann 1998; Stein 1996). Some scholars also demonstrate a range of possible relationships between producers and elite consumers (Brumfiel 1998; Costin 1998a, 1998b; Sinopoli 1998; Underhill 2002; Wright 1998). Stein (1996:25-26) explicitly points out, "the distinction between attached and independent specialists should not be viewed as an absolute dichotomy, but rather as a characterization of two different ends of a continuum of variation in the degree of elite control over craft production". Perhaps it may be said that there are a variety of attached specialists. I will attempt to demonstrate the complexity in the social context of production in my study.

Attached specialisation is a special phenomenon of complex societies, as a political-economic strategy of the elite. However, it does not necessarily appear with the emergence of complex societies. It could occur in a chiefdom society (Earle 1987), but it is more likely to be associated with state societies (Stein 1996; Underhill 2002). In complex societies, especially state societies, there can be "dual" craft economy: both independent and attached specialists operate together in a single society (Stein 1996:32; Stein and Blackman 1993:53). On the other hand, if specialised craft production appears in a simple/egalitarian society as the result of extra-household divisions of labour, as discussed in the previous section, it is possible that such specialists are not completely independent in deciding production and distribution, but rather dependent on their kin groups. Certainly, this kind of "attached" or "subsidiary" specialisation in simple societies is qualitatively different from attached specialisation in complex societies in terms of product style, artisan's identity, social function, distribution, and so on.

As discussed above, independent specialists usually produce utilitarian goods for a vast range of social demands, and attached specialists tend to manufacture valuable prestige goods for exclusive political requirements of elites. Ceramic wares were not an important kind of prestige good in many areas of the world, and production and use of pottery vessels often are regarded as peripheral to the development of social stratification (Underhill 2002:8). Therefore, pottery production is mostly conducted by independent specialists, and is not under the direct control and administration of elites or centralised institutions; this can be exemplified by the early state societies of mid-third millennium B.C. Mesopotamia (Stein and Blackman 1993; Wattenmaker 1998). However, the interrelationship between the type of products and the context of production does not always correspond that neatly. Some scholars question the necessary correlation between product type and production context, and suggest that the emphasis of context studies should move away from the products toward the manner by which the producers are compensated for their work (Flad 2004:55). In fact, attached production may involve utilitarian wares as a revenue-generating activity for elites, and independent production may also involve high-labour luxury items in societies where there are no sumptuary restrictions (Costin 1991:38). There are indeed some archaeological cases in which attached specialists produce utilitarian goods, including utilitarian pottery wares. For example, in Late Uruk Mesopotamia, the dual craft economy consisted of independent and attached production. It is probable that only those high-value prestige goods were completely under the control and monopoly of the elite or governmental institutions, whereas other low-status goods including utilitarian ceramics would have been manufactured both by independent and attached specialists (Stein 1996:32). In early dynastic Egypt, the government or political leaders also controlled production of utilitarian goods, including mass-produced pottery (Wenke 1991:307-308). In the Inka Empire, some local potters were mobilised periodically to manufacture Inka-style wares in state-sponsored production facilities. The huge quantities of state-style ceramics were found throughout the empire as local-style pottery. Only the finest Inka-style wares, especially some plates and *keros*, were more exclusively used in elite feasts, though they were not the highest prestige goods yet (Costin 1996). These cases exhibit the complexity of production context once again. My study will also demonstrate the particular relations between "attached production" and utilitarian goods.

4.3. Approaches to Identifying the Mode of Production or Specialisation

4.3.1. Direct Evidence

The mode of production (or the organisation of production) can be reconstructed through direct and indirect evidence of production.

Costin (1991:18) argues that "The most straightforward way to identify the organisation of production is to locate places at which production took place". She (1991:19) and many other scholars (e.g., Feinman et al. 1984:308-309; Stark 1985) present similar schemes in this aspect. Direct evidence linked to locations of pottery production basically involves four elements: 1) structures in relation to manufacture or storage, 2) facilities associated with production, such as kilns or firing pits, 3) pottery-making tools and raw materials, such as turntables, wheels, anvils, moulds, scrappers, paddles, un-worked clay, pigments, and so on, and 4) refuse or kiln wasters (i.e., misfired shards).

On the basis of such direct evidence, the mode of production can be evaluated by identifying productive context, concentration, scale, and intensity. For the context of production, attached production is often physically linked to elite domestic structures or governmental facilities, whereas independent production is usually associated with commoner domestic architecture or separate workshops. In addition, the existence of prestige goods usually indicates attached production too. The concentration of production can be identified through distribution of production debris or production loci among communities in a region. The scale of production can be inferred from production settings (domestic space or separate workshops) or the size of production facilities. The intensity of production is the most difficult parameter to identify in archaeological record. For household-based production, the intensity of production can be inferred from the range of economic activities reflected in the catalogue of tools used by households, but this way is ineffective for non-household-based production (Costin 1991:25-32).

Nonetheless, the intensity of production, as well as the other three parameters, can be further reflected by indirect evidence of production.

4.3.2. Indirect Evidence

Because direct evidence of the ancient organisation of production is not always preserved, and is not easily discovered, indirect evidence is more often employed by archaeologists. "Indirect data are recorded from the finished objects themselves, rather than from the features and artifacts associated with their production" (Costin 1991:32).

Indirect data involve standardisation of items, efficiency and skill in manufacture, regional variations of particular artefact types, and so forth (Costin 1991:32-43; Costin and Hagstrum 1995). Among them, standardisation is one of the most effective variables for assessing the degree of specialisation, and is most frequently used by archaeologists.

Among the four parameters assessing the mode of production, the intensity of production can be most effectively investigated through standardisation index of dimensional variables of wares, especially for utilitarian goods. Most scholars contend that the intensity of production, which can be reflected in artisans' skill, experience, and talent, is usually positively correlated with ceramic standardisation (e.g., Costin and Hagstrum 1995; Longacre et al. 1988; Roux 2003; Underhill 2003).

Standardisation can be measured through diverse aspects, such as raw material composition and manufacturing techniques, form and dimensions, and surface decoration (Blackman et al. 1993; Stein and Blackman 1993). Because I could not get the data on the chemical composition of pottery, in this study I will focus on the standardisation of the other variables through analyses of nominal data and metric data.

4.3.3. Standardisation, Diversity, and Specialisation

Standardisation and diversity are one of the most significant measurements used to assess the variability of specialised production. As Rice (1991:257-258) states:

> Standardization and diversity have continued to be important in later studies, and the increased attention given to them may be seen as an outgrowth of two important trends in archaeology today. First, both concepts can be seen as strategies for understanding variability

> within archaeological data sets, for example, for seeking meaningful patterns in a heterogeneous body of material. ... Second, within the context of a study of specialization, the use of these concepts represents a strategy for overcoming the widespread lack of direct data bearing on productive organization, as well as a means to develop more effective indirect indicators. The application of ideas on ceramic standardization and diversity to problems of socioeconomic organization may be seen as efforts to develop "middle-range theory" or "bridging argument" between hard data and abstract theory.

Rice (1991:268) defines "standardisation" as this:

> Standardization refers to a relative degree of homogeneity or reduction in variability in the characteristics of the pottery or to the process of achieving that relative homogeneity. The important term here is "relative". There is no single scale, no decontextualized measure, no quantitative index with an absolute zero for comparing variability and standardization of pottery through time and space. Standardization is not a matter of presence/absence but one of degree.

Similar to Rice, some other scholars also stress the "relativity" of standardisation, and postulate that standardisation should be recognised in comparison (e.g., Arnold and Nieves 1992:94; Costin 2001:302).

What are factors responsible for standardisation? In the light of their ethnographic research, Arnold and Nieves (1992) suggest that the intended market, the fabrication technique, and the potters' subjective perceptions of the variability of particular vessel shapes may affect the variability or standardisation in ceramic products. Costin (1991:33-34) similarly attributes standardisation to efficiency, consumers' demand (market), and distributors or government regulation. In addition, most scholars argue that standardisation is also associated with producers' experience and skill, or scale and intensity (from part-time to full-time) of production (e.g., Frankel 1988:34-35; Longacre 1999; Longacre et al. 1988; Roux 2003; Sinopoli 1988; Underhill 2003). In fact, the direct reason resulting in standardisation is few producers and routinised or "industrialised" production. Therefore, in essence, standardisation should be the result of specialisation (Costin 1991:33; Sinopoli 1988:582; Stein and Blackman 1993:31). In other words, standardisation implies specialisation. "The underlying assumption is that a high degree of standardization or homogeneity in vessel dimensions reflects specialized mass production" (Blackman et al. 1993:61). However, it is not in turn appropriate to equate specialisation to standardisation. Specialisation may also be embodied in a variety of other ways, such as elaboration and diversity (Rice 1981, 1984, 1991).

The most elaborate definition of "diversity" still comes from Rice (1987:202-203):

> Diversity is a concept borrowed from population ecology, ...In archaeological assemblages, diversity means variability not in the properties of individual sherds but in the numbers and frequencies of different categories of pottery, which may be taxonomic units, form classes, decorative styles, and so forth. Diversity is a statistical measure akin to "variance," which summarizes the kind and amount of variation around a mean; in this case the "mean" is the constitution of an assemblage. The variation is described in terms of two components: richness, or the number of categories present, and evenness, the distribution of individuals within the categories.

High diversity in a ceramic assemblage indicates many producers, many kinds of products serving distinct needs, a lack of controls or regulation of manufacture, or some combination of these factors. Conversely, low diversity reflects fewer producers, fewer kinds of products (or more products, each being less variable), a controlled or regulated manufacturing process, or some combination of these factors, and characterises the specialised production in complex societies for the mass production, repetition, quality control, and more (Rice 1991:274). However, this kind of interrelationship between diversity and specialisation is more applicable to utilitarian vessels. For prestige wares, high diversity, especially in shape and decoration/style classes, usually reflects some kind of specialist production (Rice 1981). Diverse prestige wares can satisfy the needs of ritual and feasts manipulated by the elite.

In fact, the interrelationship between diversity and specialisation is not entirely definitive. According to Costin, gross formal and stylistic diversity within an assemblage is not the best analytical measure for studies of specialisation for two reasons: "First, both large and small concerns, specialized or unspecialized, are equally capable of producing a variety of stylistically or formally different wares. Second, variability at this level often reflects sociopolitical processes or functional concerns, rather than economic organization" (Costin 1991:34-35). Therefore, the significance of diversity should be interpreted in combination with function of pottery and specific social context.

4.3.4. Measurements of Standardisation, Diversity and Variation

Ceramic standardisation and variability can be investigated through five broad attribute systems: manufacturing techniques, composition or texture of paste, colour, surface treatment and decoration, and form (primary and secondary shape characteristics, dimensions). The variation in these attribute elements can reflect the tendency of diversity and/or standardisation. The five elements include nominal and metric variables. Diversity can be evaluated in the light of richness and evenness of nominal variables in the five trait categories, and standardisation can be measured through statistical analyses of metric variables (Costin 1991; Rice 1991).

The method most often used in assessing metric variables is descriptive statistics, such as mean, standard deviation, and the coefficient of variation (e.g., Allen 1992; Benco 1988; Blackman et al. 1993; Costin and Hagstrum 1995; Frankel 1988; Longacre 1999; Longacre et al. 1988; Roux 2003; Stein and Blackman 1993; Underhill 2003). The coefficient of variation (CV) describes relative variation by expressing the standard deviation as a percentage of the mean, thereby removing scale effects (Longacre et al. 1988:103). Like most scholars, I will focus on CV values for the analyses of standardisation/variability of metric data of pottery in this study.

All measurements of metric variables are based on taxonomy of pottery, which involves primary and secondary formal classes. Traditional pottery typology is still a useful method, through which we can establish chronological sequences and shape classes of pottery which are the very important bases for measuring, recording and analysing ceramics.

I will investigate variation of the mode of production between four cultural periods or assemblages from the Neolithic to the early Bronze Age in the Yuanqu Basin (see next section). Almost every period can be subdivided into two or three phases. For assessing inter-group variation of two or more samples, the most effective and direct method is the analysis of variance (ANOVA, F-ratio), a kind of inferential statistics (i.e., Blackman et al. 1993; Longacre 1999; Longacre et al. 1988; Stein and Blackman 1993; Underhill 2003).

The analysis of variance technique involves the estimation of two independent measures of variance—one based on the variability between the groups and the other based on the variability within groups. The F-ratio provides a measure for assessing the difference between two variances. A large F-ratio implies that differences between the groups are greater than differences within any groups, and therefore the differences between the groups may be statistically significant. Conversely, a small F-ratio may imply that there is no statistically significant differences between the groups (Sinopoli 1991:193-194).

I will use ANOVA to evaluate variations between chronological phases within each period. However, this method is not appropriate for assessing variations between the periods/assemblages in this study, because the assemblage composition of pottery classes among most periods is almost completely different from one another.

4.4. Data Collection

4.4.1. Sources of Ceramic Data

The range and contents of direct archaeological evidence of pottery production have been discussed above. The direct evidence in the Yuanqu Basin mainly encompasses some kilns, storage pits, paddles and anvils for pottery manufacture, known both from published reports and unpublished records. The latter is available in the workstation of the Gucheng archaeological team in Yuanqu. Here I mainly detail the collection of indirect data—pottery itself.

It is necessary to point out firstly that the ceramic objects involved in my study only include the ceramic containers and a small number of ceramic tools used for pottery manufacture. The ceramic tools for other productive activities are not considered.

As mentioned in Chapter Three, there are 46 sites known in the Yuanqu Basin, seven of which have been excavated. For the surveyed sites, we only have a small number of sherds, which provide little information for the reconstruction of modes of production. Of seven excavated sites, Dongguan and Nanguan, which were excavated by our archaeological team (Gucheng Team), are the only two sites whose ceramic data are completely available for me. Dongguan contains the remains of the Early, Middle and Late Yangshao, Miaodigou II, and Longshan period. Nanguan includes the remains of the Late Yangshao, Erlitou, and Erligang period. The earliest remnants dating to the Peiligang period in the Basin were not present in the two sites. Therefore, these two sites contain the remains of every period from the Early Yangshao to the Erligang period, namely from the Late Neolithic to the Early Bronze Age (ca. 4600-1300 B.C.). However, for the Middle and Late Yangshao and the Longshan period, the number of ceramic vessels is comparatively small, and not enough for the quantitative analyses of ceramic samples. Accordingly, I choose four substantial assemblages as the main focus of the pottery study. These four assemblages come from the Early Yangshao and Miaodigou II periods of the Dongguan site, and the Late Erlitou and Erligang periods of the Nanguan site. Fortunately, these four periods cover the social changes from simple to complex society, namely from local groups/tribal societies to the regional polities, including chiefdom and state societies (see Chapter Three). These diachronic changes are necessary for the investigation of relationships between productive modes and social forms. Moreover, importantly, these four assemblages have good ceramic samples, from sites with well-identified socio-political forms.

Organisation of pottery production can be investigated archaeologically through direct and indirect evidence, but the latter frequently constitutes the main body of data because direct evidence is relatively rare in archaeological sites. In my study, the changes of modes of production over time will be investigated by the comparison of different periods. Two critical issues need to be resolved as the preconditions guaranteeing the reasonable and effective analyses of pottery production. One is the establishment of analytical units, including spatial and temporal units. The other is the exact classification of ceramic vessels.

4.4.2. Establishment of Analytical Units

Many crucial insights used in pottery study are derived from ethno-archaeology. The advantage of ethno-archaeology is that almost every step and detail of pottery production, distribution and consumption can be observed directly, including organisations of production. In contrast, the disadvantage of archaeology is obvious, because the reconstruction of pottery economic systems can only rely on inferences from material data rather than direct observation of on-going activities. In ethno-archaeology, the analytical units may be composed of individual potters, workshops, or productive seasons (i.e., Longacre 1999; Roux 2003; Stark 1995b; Underhill 2003). However, it is extremely difficult to identify the products of individual potters, workshops, or short-term productive events in archaeology. Any one analytical unit in archaeology usually includes products from a large spatial area and temporal span.

The primary analytical units in this study are ceramic assemblages of the four periods mentioned above. Statistical comparability between assemblages depends on approximately equal spatial units and time span (Rice 1991:276). These four assemblages consist of a group of random samples of the specific cultural period recovered in a single site (Dongguan and Nanguan), so they have equivalent geographical sources. They are also comparable in time span.

The Early Yangshao period of the Dongguan site lasted approximately from 4600 B.C. to 4100 B.C. The remnants of this period can be divided into two distinctive cultural types: Dongguan I and Dongguan II (see Chapter Three). There are only a very few ceramic vessels discovered in Dongguan II (National Museum of Chinese History et al. 2001), so the Early Yangshao ceramics used in this study only come from Dongguan I (ca. 4600-4200 B.C.). The time span of Miaodigou II at the Dongguan site was about 400 years (3900-3500 B.C.). The Late Erlitou and Erligang respectively spanned 200 years (1800/1700-1600/1500 B.C.) and 300 years (1600-1300 B.C.).

Every cultural period can be subdivided into two or more chronological phases (see Table 4-1). The excavation report of Dongguan partitions Dongguan I into three phases (National Museum of Chinese History et al. 2001:479-483), but this scheme of chronological division is problematic. I prefer to divide the Dongguan I period into two phases: the Early and Late phase, each of which spanned about 200 years. The Miaodigou II period at Dongguan is more appropriately subdivided into three phases: the Early, Middle and Late phase (National Museum of Chinese History et al. 2001), and the average time span of each phase was more than 130 years. The Late Erlitou Period spanned about 200 years, and it is one phase of the entire Erlitou period (consisting of four phases or Early and Late phases, see Table 1-1 and 1-2), and cannot be subdivided at Nanguan. The Erligang period may be divided into two phases: Lower Erligang and Upper Erligang, each of which spanned about 150 years. The time span of the phases within every period is approximately equal with little variation between them, just from 130 to 200 years, which should be an acceptable variation for the long-term comparison in archaeology. These phases within every cultural period constitute a finer scale of secondary analytical units. The comparison between ceramic assemblages will be carried out primarily at this finer scale of the secondary analytical units (phases) within every period, and then between the four primary analytical units (periods).

Table 4-1. Chronological phases and time span for the four periods.

Period	Phases	Time Span (ca. years)
Erligang	Upper Lower	150 150
Erlitou	Late Early	200 200
Miaodigou II	Late Middle Early	130 130 130
Early Yangshao (Dongguan I)	Late Early	200 200

4.4.3. Classification of Ceramic Vessels

"Classification is a tool that allows us to group individual ceramic fragments—whether shards or whole vessels—into coherent and meaningful groups whose members share some essential characteristics" (Sinopoli 2002:64). Classifications of pottery can be developed from diverse aspects, such as function (storage, preparation, cooking, serving, etc.), social identity (utilitarian and prestige goods), paste or colour, shape classes, and so forth. Although all these aspects will be addressed in this study, the classification discussed here mainly refers to differentiation of ceramic classes or ceramic types, namely shape classes. Inaccurate classification of ceramic categories can directly result in the negative impact to pottery analyses, especially to values of descriptive statistics (Costin and Hagstrum 1995:631-632; Longacre et al. 1988).

The most ideal classification of ceramic classes should be consistent with actual classification by ancient potters or residents, but it is difficult, if not impossible, to reach this point in archaeology. Classification is comparatively easy in ethno-archaeology. Ethno-archaeologists can even observe and record distinctive products of individual potters within a single ceramic type. However, it is very difficult to establish ceramic groups of individual potters in archaeology (Frankel 1988:49).

Despite facing these difficulties, archaeologists can still make classifications of ceramics significant. Just as Sinopoli (2002:64) points out, "by basing our classifications on patterned and verifiable variations in archaeological materials, we can encode categories that we believe were in some way significant to the past makers and users of ceramic vessels".

One of the important principles for classifications of ceramics is that differentiation of all ceramic categories should comply with the common standards. In this study, I categorise ceramics principally in terms of vessel form and size. According to these standards, some primary shape classes can be subdivided into two or more secondary shape classes (subclasses). For some ceramic classes, the differentiation of size classes is very important (Roux 2003:778; Stark 1995a). If one category of ceramics subsumes several size classes unrecognised by archaeologists, the variation statistics calculated on this ceramic category will be misleading.

In this study, my classification of pottery is based on the taxonomy in the excavation reports of Dongguan and Nanguan (National Museum of Chinese History et al. 1996, 2001), but is not same as the reports. I have re-classified all the available ceramic vessels, and modified some classifications in the excavation reports. The ceramic classifications of every period in this study will be specified in the following chapters.

4.4.4. Measurement and Recording of Pottery

All ceramic specimens of the Dongguan and Nanguan site are stored in the specimen rooms of our archaeological team in Yuanqu. These samples include all whole and recovered vessels, and a vast majority of sherds whose shape classes can be identified. Therefore, the quantity of specimens of each ceramic type can be regarded as a random sample in a specific assemblage, and the variations in proportion among ceramic types can be taken to represent the original proportion of ceramic types in the assemblage. This point is very important for the comparison of intra- and inter-assemblages in constitution of ceramic classes.

Every assemblage contains many ceramic classes. For most of them, the sample size is too small to be analysed quantitatively, so I selected those classes with relatively large sample size from every assemblage to measure and record for the statistical analyses. The total number of specimens I measured is nearly 2000. I later excluded several ceramic types whose distributions among chronological phases within an assemblage are extremely unbalanced. The actual number of ceramic specimens used for statistical analyses is 1353.

After establishing categories of pottery, the next step of data collection is measuring and recording diverse variables of pottery, which include metric and nominal variables, namely quantitative and qualitative data.

Metric variables include orifice diameter, neck diameter, body width, base diameter, rim length, and vessel height. For tripods, I also measured leg length and body height. Figure 4-1 illustrates these measurements for two types of vessels. In addition to these single dimensions, I will also assess variation in the ratio, which focuses on the ratio of orifice diameter-to-vessel height in this study. "Using ratios, the size problem vanishes and we become concerned only with variation in the ratios" (Longacre et al. 1988:109). Accordingly, "ratios might provide better tests for degree of vessel uniformity than single dimensions" (Underhill 2003:247).

Nominal variables include techniques of pottery forming and finishing, texture of ceramic paste, colour of ceramic surface, as well as surface treatment and decoration.

Forming techniques mainly comprise coiling, moulding and throwing. Some categories of ceramics are manufactured by combination of these techniques. In addition, so-called "slow wheel" (turntable) technique was increasingly applied to shaping and smoothing rims and bodies of ceramic vessels in the Yuanqu Basin. The result of this technique was that clear "rilling"—rhythmic ridges and grooves spiralling around the vessel walls (Rice 1987:129)—was left on the surface of rims and bodies, but it is not the marks of wheel throwing.

The analyses of mineralogical and chemical composition of ceramic paste are essential measures for identifying provenience of pottery. The most widely used method in these aspects is X-ray spectroscopy analysis or instrumental neutron activation analysis (INAA) (e.g., Arnold et al. 1999; Blackman et al. 1993; Stein and Blackman 1993). Regrettably, I could not apply these techniques because of the limitation of budget cost and fieldwork time. I therefore recorded the texture of ceramic paste, mainly referring to types of inclusions or temper visible in cross section of vessels. The main tools for this work included a low-level magnifier and a geological sand grain chart for identifying types of inclusions. The sand grain chart separates six types of fabrics:

Very coarse sand	1.0-2.0 mm
Coarse sand	1/2-1.0 mm
Medium sand	1/4-1/2 mm
Fine sand	1/8-1/4 mm
Very fine sand	1/16-1/8 mm
Silt	<1/16 mm

Sand is the major inclusion or temper of ceramics in the Yuanqu Basin assemblages. These six fabrics can nearly cover all types of ceramic texture for every assemblage. Actually, it is almost impossible to distinguish fine sand from very fine sand in ceramic paste in practical observation with a low-level magnifier, so I regard them as one category of inclusion and use "fine sand" as the name. Hence the list of fabrics I employed just comprised five types of six fabrics. In most situations, inclusions from very coarse sand to fine sand may have meant temper—deliberately added inclusions in ceramic paste. For silt, no sand grain is visible under a low-level magnifier, and I think that there is no temper for the silt paste of pottery. In brief, the texture of ceramic

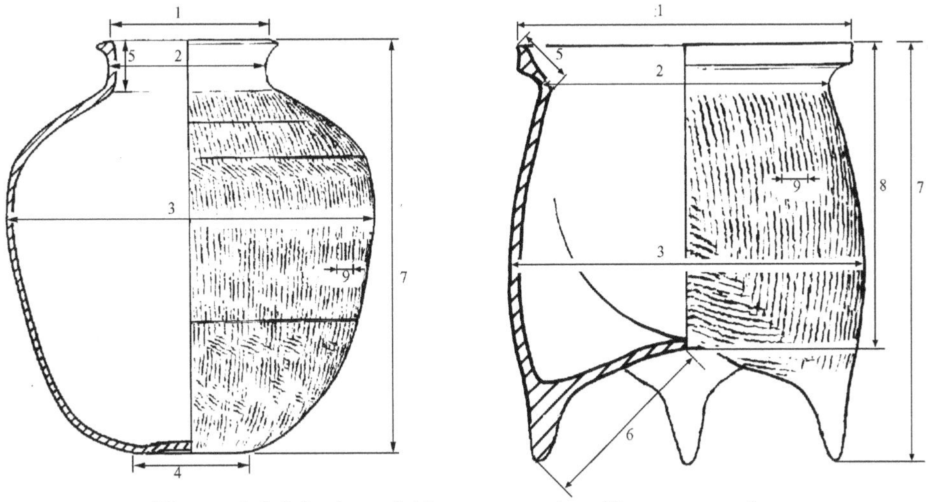

Figure 4-1. Metric variables measured on Yuanqu vessels.
1. Orifice Diameter 2. Neck Diameter 3. Body Width 4. Base Diameter 5. Rim Length 6. Leg Length 7. Vessel Height 8. Body Height 9. Width of five marks

paste is a significant variable for comparing standardisation and diversity between assemblages.

The identification of ceramic colour is often difficult in the Yuanqu Basin assemblages, especially for the pottery of Miaodigou II and Erlitou & Erligang, because a lot of vessels were affected by environments of use or deposition, and their original colours were changed. Many complete vessels have two or more colours on the surface, and even on the interior body. In some circumstances, two sherds that can be refitted together display two completely different colours. Most ceramic samples discovered in the Yaunqu Basin are fragmentary sherds rather than whole vessels, so I often could not identify the original colours of vessels merely through these sherds. Though I recorded the surface colour of most specimens according to the main body of sherds or whole vessels (in some circumstances it is easy to identify ceramic colours), I do not intend to conduct statistical analyses, but only assess the general trends in the colour between assemblages.

Surface treatment and decoration of pottery involved in this study may be grouped into three major categories: plain (*su mian*, ranging from lightly smoothed to highly burnished), cord-marked (*sheng wen*), and basket-marked (*lan wen*). Other surface treatments comprise a small quantity of check stamping (*fangge wen*) and bowstrings (*xian wen*, horizontal, circumferentially-incised or engraved lines) (see Railey 1999:241-242). Besides, clay strips (*fujia duiwen*) were also a sort of popular decoration, and were usually used as attached secondary parts in combination with the main decorations.

The techniques of cord-marking include cord-wrapped paddle-impression and corded dowel-impression, whereas a vast majority of basket-marking is executed with basket-engraved paddles. For paddle-impressed cord-marking and basket-marking, some recurrent design units can be discerned in some circumstances. I tried to measure and record the width and the number of marks of these designs. Unfortunately, there are only a very small number of vessels whose design units can be discerned because most samples are very small sherds and lack complete designs or their designs are ambiguous. Therefore statistical analyses are not feasible for the comparison of these two variables of designs between assemblages. I adopt an alternative approach to deal with this. I specified five marks as one measured unit that are available for most samples, and then I measured and recorded the width of one unit (a group of five marks) for any measurable vessel (see Fig. 4-1). Although there is a little individual variation for the width of marks on most vessels, the width of five marks is basically identical for every such unit on a single vessel. The data in this aspect will be analysed statistically as a component of metric variables.

4.4.5. Sample Size

The sample size should be adequate for statistical analyses. Although there is not an absolute criterion for sample size in measurements of standardisation and diversity, Rice suggests at least 30 individuals as a minimum sample size for a data set (Rice 1989:116). Here I try to reassess the minimum sample size through a group of vessels with a large sample that will be addressed in this study. Figure 4-2 shows the distribution and variation of CV values (coefficient of variance) with different sample sizes for the orifice diameter of *li* tripods in the Lower Erligang period. Each sample is selected randomly from the same set of 90 individuals. The result indicates that 20 seem to be a pivotal point. The variance curve of CV values after that point could keep at a relatively stable level though slight fluctuations may still appear. Accordingly, I would view 20 individuals as the minimum sample size, when we do statistical analyses for a data set under the optimal condition.

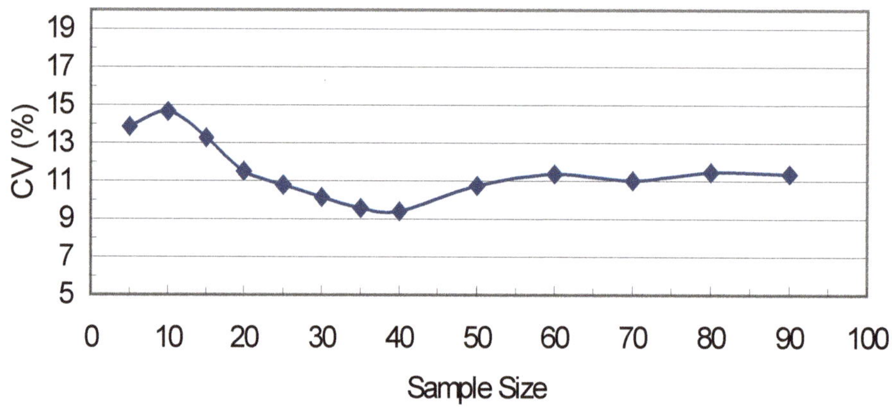

Figure 4-2. Distribution of CV values with different sample sizes for the orifice diameter of *li* tripods in the Lower Erligang period at Nanguan.

However, this condition cannot always be met in archaeological records. Sample size for any one group of ceramic vessels from archaeological sites does not depend on our subjective wishes, but relies on practical archaeological discoveries. Sometimes archaeologists have to deal with smaller samples and evaluate their statistical significance (e.g., Allen 1992; Blackman et al. 1993; Stein and Blackman 1993; Underhill 2002). Even in some ethno-archaeological cases, samples of some data sets are also less than 20 (i.e., Roux 2003).

Though I measured and recorded more than 1000 pottery specimens in the Yaunqu Basin, which are almost all the vessels available for statistical analyses, they come from a large number of ceramic types from different phases of the four periods (assemblages), so the number of samples for some ceramic classes in a chronological phase is less than 20. Since many ceramic classes mainly consist of sherds rather than whole vessels, most variables within a group of vessels are usually unbalanced and unequal in the number of specimens, and some variables keep at a lower level in sample size. In this thesis I preserve the statistical data whose number of specimens are more than five, and abandon those statistical data whose number of specimens are five or less.

I have noted that samples of some data sets for specific phases are very small, and likely results in deviation of statistical values. This situation is beyond my expectations for sample size before the fieldwork. On the other hand, there are still many data sets more than 20 in sample size, especially in the Miaodigou II and Erlitou & Erligang assemblages. These data sets compose the basis of statistical analyses in this study. I will assess significance of the statistical values whose samples are less than 20 in number by comparison with values of those more than 20 in sample size.

The final aim of the pottery studies in this thesis is to provide insights into variation of modes of production between the four cultural periods, so I will eventually compare statistical data between the four assemblages, using pooled data of each ceramic type from different phases within every assemblage. Fortunately, a great majority of pooled data for the four assemblages are more than 20 in sample size.

4.5. Summary

The main purpose of the pottery studies in this dissertation is to investigate the evolution of modes of pottery production from simple to complex societies in the Yuanqu Basin. This chapter discusses some crucial theoretical and methodological models concerning the mode of production and specialisation, and details the critical methods and pathways of collecting and analysing pottery data.

For the past two decades, the relationship between craft specialisation and social complexity has been a focus of concern and controversy among a number of archaeologists and anthropologists. Before exploring this issue, however, we need to first identify diverse types of specialisation, and then investigate interrelationships between the variability of specialisation and social forms.

The identification of productive organisations will be mainly based on four modes of production suggested by Peacock (Peacock 1982) and Rice (Rice 1987): *household production, household industry, individual workshop industry,* and *nucleated workshops*. As it can be difficult to identify specific forms of organisation of production in archaeology, I will assess the changes of productive organisations and degrees of specialisation on the basis of four parameters established in Costin's model: the *context, concentration, scale,* and *intensity* of production (Costin 1991). These four parameters can be investigated through direct and indirect evidence of pottery production. Direct evidence involves those visible facilities, tools and wasters linked to pottery production in archaeological sites. Indirect evidence includes nominal and metric data of ceramic vessels, through which standardisation, diversity, and specialisation can be evaluated. Descriptive and inferential statistics will play critical roles in analyses of metric variables of pottery.

Four cultural periods abundant in pottery in the Dongguan and Nanguan sites are investigated. These four periods (Dongguan I, Miaodigou II, Late Erlitou, and

Erligang) represent distinct social forms from simple to complex societies, and constitute the primary analytical units in this study. Most periods can be subdivided into two or three chronological phases, which constitute secondary analytical units.

Comparison among periods and phases will be carried out through diverse ceramic categories. I choose some major ceramic classes in every assemblage as the basic analytical elements. Combining the indirect evidence with direct evidence of pottery production, I will first assess variation between the phases within each period, and finally reconstruct the process of development of modes of pottery production in the four periods, and investigate relationships between productive organisations and social forms.

Chapter Five

Pottery Production in a Tribal-level Society: A Case Study of the Early Yangshao Period at the Dongguan Site

5.1. Introduction

The early Yangshao period in the Yuanqu Basin (ca. 4600-4100 B.C.) encompassed two cultural types: Dongguan I and Dongguan II. As mentioned in the previous chapter, this study only addresses Dongguan I (ca. 4600-4200 B.C.).

As discussed in Chapter Three, the social form of the Early Yangshao period in the Basin was characterised by the local group/tribal society. This chapter focuses on the mode of pottery production during this period. The main body of data is from the assemblage of Dongguan I at the Dongguan site.

The excavation report of Dongguan further partitions Dongguan I into three phases according to the assemblage constitution of ceramic classes and attributes of vessel forms (National Museum of Chinese History et al. 2001:479-483), but actually there is little difference between the first two groups of ceramics, so I have incorporated the first two phases into one phase, and then subdivide the whole Dongguan I period into two chronological phases: the Early and Late phase, each of which spanned about 200 years.

Based on the theoretical and methodological models outlined in last chapter, the following studies will try to assess the mode of production during the Dongguan I period from *context*, *concentration*, *scale*, and *intensity* of pottery production, through direct and indirect evidence. The analyses of direct evidence will address a few contemporary sites beyond Yuanqu to help understand the circumstance of Dongguan. The analyses of indirect evidence will include nominal and metric variables of pottery.

5.2. Direct Evidence of Pottery Production

The most visible evidence of pottery production consists of structures, facilities, tools, raw materials, and wasters (see Chapter Four), but these are seldom preserved, or are not easy to discover in archaeological excavations. At best, we can only find some scattered evidence of them.

Direct evidence of pottery production is not abundant during the Dongguan I period at the Dongguan site, but we have got some critical clues. For pottery-making tools, only one anvil was discovered in an ash pit, H126, belonging to the late phase (Fig. 5-1). In addition, more importantly, we found three pottery kilns of Dongguan I during the surveys in 2001. These three kilns are exposed at the eastern bluff edges at Area IV. Two of them are close together in the north, and the other is about 160 m south of them (Fig. 5-2). Judging from the surviving remnants of these kilns, they are all very small, probably measuring around one meter in diameter. The associated surface collections of small sherds do not allow them to be separated chronologically, so they are regarded as in use at about the same time. Because all these kilns are located at bluff edges, it is likely that there are other undiscovered kilns near them inside the bluffs, and that some have been destroyed with the erosion of the terrace outside the current bluffs.

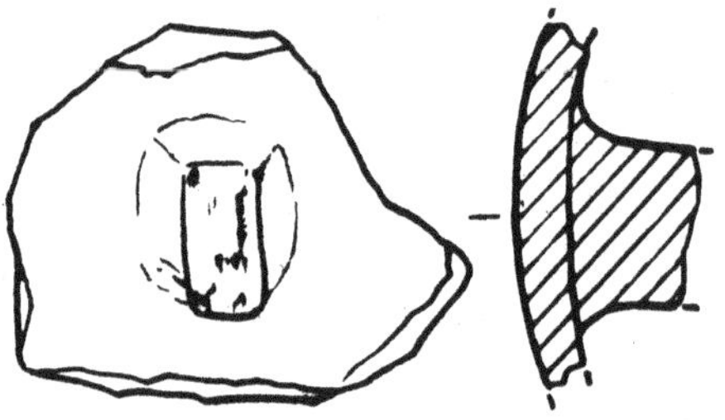

Figure 5-1. Anvil of the Dongguan I period discovered in the Dongguan site.

This evidence suggests that a large quantity of pottery (if not all) was made within the community during this period in the Dongguan settlement. It is unclear if pottery production was conducted at the other four contemporary settlements (see Chapter Three) as there was no direct evidence discovered through surface surveys.

As mentioned in Chapter Three, the Dongguan I component was limited to the northern part of the Dongguan site—Area IV. The three exposed kilns and other possibly undiscovered kilns were situated at the eastern edge of this area, outside the residential areas. This may have been the main area for pottery firing. With regard to locations of pottery manufacture, one possibility is that pottery was mainly fabricated within residential zones, and then moved to the special firing area. The other possibility is that pottery manufacture was mainly carried out near the kilns. To the east of the bluffs is the bank of the Yun River, so the pottery production here could benefit from the convenient water source (see Fig. 5-2). In addition, the excavations in Area IV showed no evidence of pottery production inside and near the uncovered houses. If a discrete area reserved for manufacturing pottery is relevant, we can argue that individual households did not manufacture pottery for their own use during this period at Dongguan.

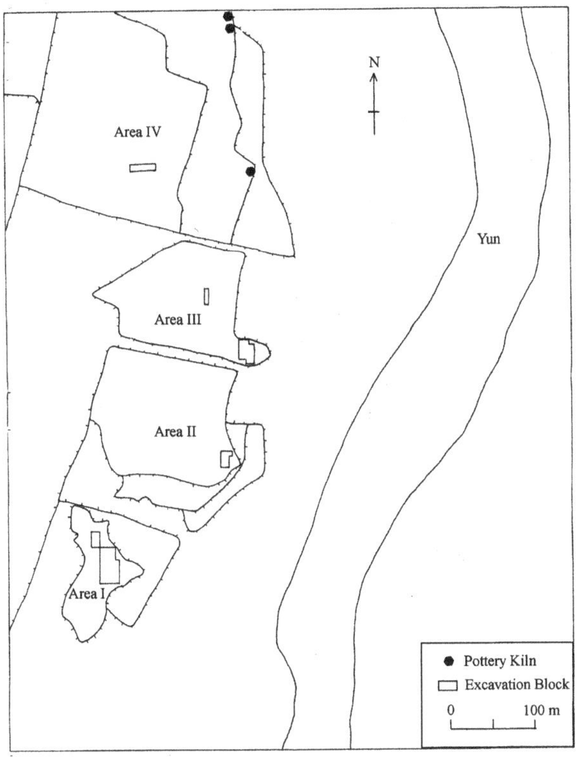

Figure 5-2. Distribution of Dongguan I kilns at the Dongguan site (redrawn from National Museum of Chinese History et al. 2001:fig.2).

The other noticeable aspect of kiln distribution is that they are separate, but not close together. The kilns cluster into two groups that respectively lie at the north and south of the settlement. This situation may have implied that at least two, if not more, units of pottery production existed in this community.

These distribution patterns of kilns and the recovery of a few residences at Dongguan can be taken to exclude the simplest mode of household production, in which pottery is occasionally manufactured in domestic context primarily for household own demands, and permanent production facilities (such as kilns) are scarce (Peacock 1982:8; Pool and Santley 1992:213; Rice 1987:184). On the other hand, those dispersed small kilns and their separate production units do not conform to measures of nucleated workshops, in which many workshops aggregate together to produce pottery on a larger scale. Accordingly, it appears most likely that the household industry or individual workshop industry was the mode of production for the Dongguan I settlement, in which pottery is relatively intensively manufactured by some households or dispersed workshops for exchanges beyond households (see Chapter Four). As these two modes are similar to each other, and it is usually difficult to discern them in archaeology, they are also grouped together as "household specialised production" (Stark 1992:184).

"Workshop" is defined as a "small activity area, covering a significant proportion of a small site, where a broad spectrum of manufacturing stages was performed together" (Tosi 1984:24). Kilns do not equate to workshops, and existence of kiln itself cannot indicate presence of workshops. Therefore, judging merely from current limited evidence, it is not certain if there existed workshops at Dongguan.

Some contemporary settlements in the Wei River valley in middle Shaanxi Province, at the west of the Yuanqu Basin, are helpful in illustrating the mode of production at Dongguan. As discussed in Chapter three, these settlements were similar to Dongguan in size (not more than 5 ha) and social forms (local groups), so it can be argued that they had analogous productive modes as Dongguan. Several early Yangshao sites, including Banpo (Institute of Archaeology 1963), Beishouling (Institute of Archaeology 1983), and Jiangzhai (Xi'an Banpo Museum et al. 1988), have been excavated on a large scale.

Similar to Dongguan, a few small kilns uncovered at Banpo were also distributed outside the residential zones, at the northeast of the settlement, and isolated from the residential zones across a long ditch (Institute of Archaeology 1963; Yan 1981, 1989a). Likewise, no direct evidence of pottery production was found within residential zones either. Conversely, two of three kilns discovered in Beishouling were situated in residential zones, which were partitioned into three house groups (Fig. 5-3). The other kiln discovered near the cemetery was also close to the southern house group (Institute of Archaeology 1983; Yan 1981, 1989a). These kilns were most likely linked with some structures. This situation indicates that the pottery production at Beishouling may have been performed in domestic context. Nonetheless, the very small number of kilns indicates that pottery was not produced by all or most households. A very limited number of households or individuals were engaged in pottery production probably not only for themselves but also for the entire community or more consumers. Because there is no evidence of workshops in the archaeological records, the most possible mode of production was household industry at Beishouling.

Most of the Early Yangshao component of Jiangzhai has been revealed by archaeological excavations, so we know about the total layout of the settlement (Fig. 5-4). More than 100 houses were distributed around the central plaza. These houses may be divided into five house groups, which were regarded as five kin groups, and encircled by a circular moat (Banpo Museum et al. 1988; Yan 1981, 1989a). At the east and southeast of the residential zones three cemeteries were discovered. The distribution of kilns is similar to both patterns displayed in Banpo and Beishouling. Two kilns (Y2 and Y3) were situated in residential zones, inside the moat, and adjacent to structures of two house groups. One kiln (Y1) was located at the east of the village, outside the moat. Four kilns collectively lay at the southwest of the village, just at the bank of the Lin River. Actually, those kilns beyond the village were also very close to the residential zones, no more than ten meters away from their respectively adjacent house groups (see Fig. 5-4). Therefore, the pottery production in relation to all these kilns may have been conducted within the residential zones, and in the domestic context. Certainly, the possibility that the special production loci or workshops existed inside or

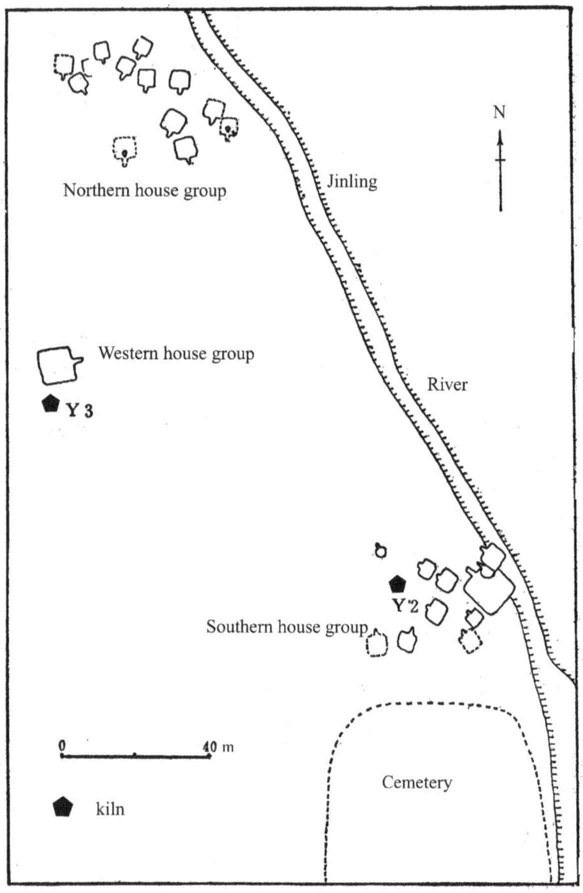

Figure 5-3. Distribution of Early Yangshao kilns at the Beishouling site (after Yan 1989a:fig.23).

outside the village cannot be excluded completely. For example, a "workshop" was found at the most southern end of the residential zones, just inside the moat. This was a small round semi-subterranean structure, measuring 2.23 m in diameter, and 0.35 m in depth. There was a platform (5-7 cm above the floor) at the north part of the structure, where there were several sets of unfired stacked bowls. Some prepared clay was placed at the south part of the structure (Banpo Museum et al. 1988:49). It is not certain if this was, in fact, a workshop because no manufacturing tools were found in the house, and it was not linked to any kiln either. The possibility that it functioned as a temporary storage room for unfired vessels and prepared clay cannot be ruled out. In a word, this was probably a special structure related to pottery production.

Regardless of productive context, it may be certain that only a small number of households or communal members in Jiangzhai produced pottery for the community. In terms of the size (around one meter in diameter), number and distribution pattern of the kilns, the most likely mode of production was also household industry or individual workshops. In addition, it is worth noting that these dispersed kilns were respectively adjacent to different house groups. This situation perhaps implied the correlation between kilns (or productive units) and kin groups.

Based on the discussion above, the distribution of kilns or production locations in Dongguan was unlikely to be different from patterns displayed in Banpo, Beishouling, and Jiangzhai, and these settlements may have been similar in the mode of production. Even

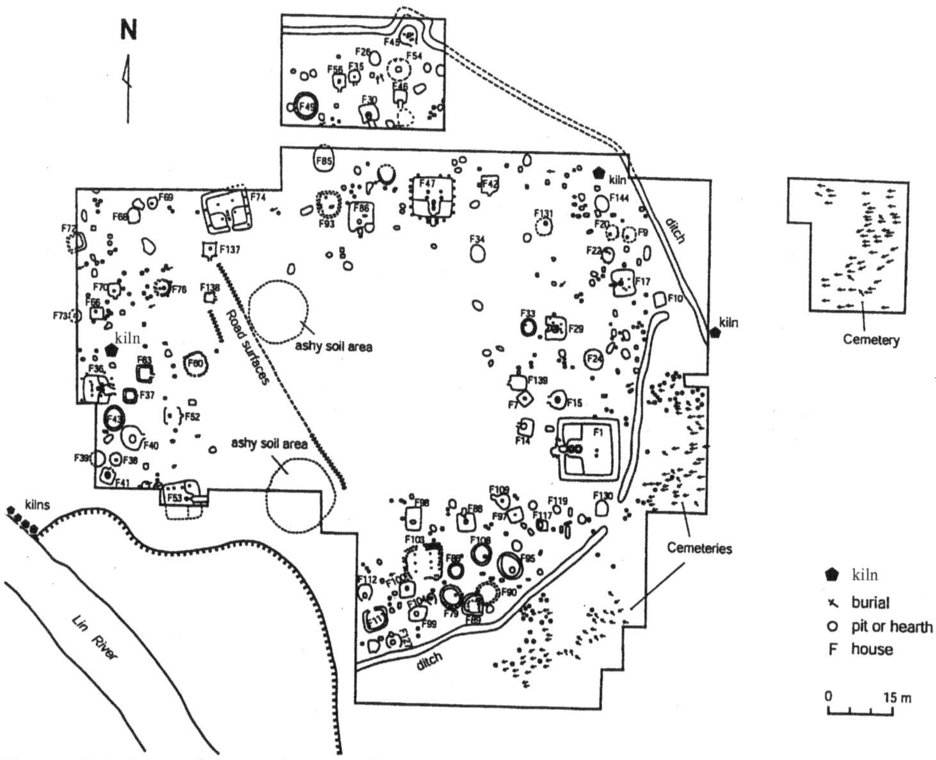

Figure 5-4. Map of the Early Yangshao settlement at Jiangzhai, showing locations of kilns (after Xi'an Banpo Museum et al. 1988:fig.6, cited from Liu 2004:fig.4.4)

though the presence of individual workshops cannot be excluded, the household industry probably was the major mode of production for this period. All three settlements that have been excavated on a large scale in the Wei River valley had kilns. This fact suggests that pottery production took place in all these settlements during this period. We can therefore speculate that in addition to Dongguan, there were other settlements in the Yuanqu Basin taking part in the pottery production. At the regional level, pottery production was probably dispersed rather than concentrated. Production may have been carried out by part-time or seasonal specialists or potters in households or scattered individual workshops. This point can be supported by the indirect evidence of pottery production (see the following analyses).

There are two possibilities concerning the social context of production and identity of producers. One is that potters and their production were completely independent, and some people or households good at pottery manufacture were engaged in production in order to increase their income, or establish a primary source of their livelihood. The other is that the pottery production was based on the division of labour within communities. The production and distribution of pottery was thus controlled and managed by communities or different kin groups, and producers were subsidiary to their communities or kin groups. This proposition fits with the suggestion by Yan (1989a) that some economic activities, including pottery production in the Early Yangshao societies, were most likely organised and managed by a certain kin groups (such as clans).

Some ethno-archaeological case studies are useful to explicate the mode of production of Dongguan I settlement and other Early Yangshao societies. In several villages in the southern part of Andhra Pradesh, India (each of them with 1200-2000 inhabitants), there are 2-5 potters. They work throughout the year according to different rhythms within the domestic space, and in most cases pottery is the only source of their income. Pots are sold within a 3-4 km radius. The firing area is generally situated at the outskirts of the village (Roux 2003:769). In San Isidro, a community of Los Tuxtlas, Mexico, a very small number of domestic potters produce pottery for a relatively restricted market, and all kilns are adjacent to the houses where potters work. These kilns are all very small, measuring around one meter in diameter (Arnold 1991). These two cases exhibit some similar production traits as Dongguan and those settlements in the Wei River valley: only a small number of potters are engaged in pottery production in domestic context within a community, and a few small kilns or firing areas are dispersed inside or outside the community. Perhaps the household industry is the most appropriate definition for their mode of production.

5.3. Pottery Classification and the Assemblage Composition

Although there are relatively obvious morphological variations for every ceramic type between two phases, the assemblage compositions of ceramic classes are also very similar between the early and late phase, so I do not intend to separately describe the classification and constitution of pottery for the two phases, but deal with them together.

There are 259 vessel specimens recorded for the Dongguan I assemblage. I identify 22 ceramic classes (including secondary shape classes) from all these samples of the two phases in terms of their shape and size attributes (Appendix 2 and Fig. 5-5). For most ceramic classes, it is easy to distinguish them by their distinctive morphological attributes. However, the differentiation of *gudibo* bowls/*gudipen* basins, and *houdibo* bowls/*houdipen* basins mainly relies on their respective vessel sizes, since there is little difference of morphological traits between them (see Fig. 5-5a). For other vessel classes, including the *pingdibo* bowl and the *qizuo* stand with relatively large sample sizes, size is not a relevant criterion.

All ceramics of Dongguan I may have been utilitarian goods. There do not appear to be any prestige goods. This situation matches the social form (the simple society) during this period.

The distribution pattern of ceramic classes in the assemblage is very diverse, but meanwhile is relatively low in evenness (Fig. 5-6). Among 22 shape classes from 259 specimens, the top five account for 57.91% of the total, ranging from 8.11% to 17.37% (including the *gudibo* bowl, *gudipen* basin, *houdipen* basin, *qizuo* stand, and *chikouguan* jar). For other ceramic types, each is less than 7% of the total (see Fig. 5-6 and Appendix 2).

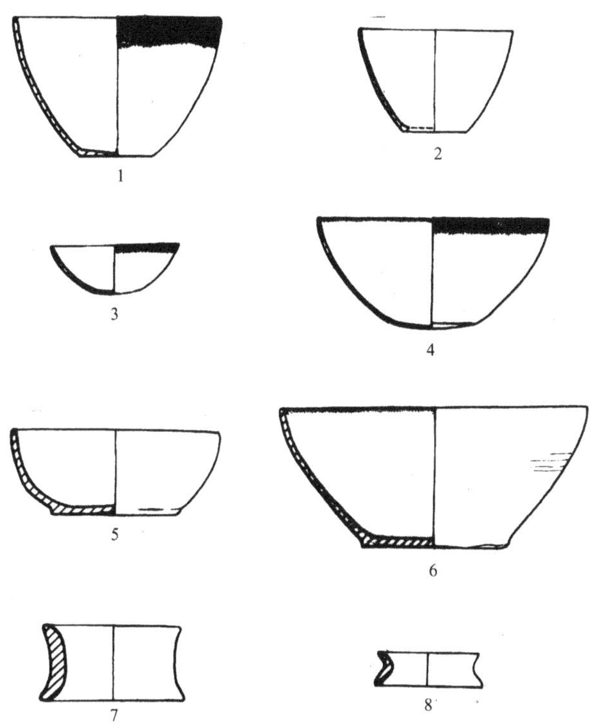

Figure 5-5a. Dongguan I ceramics from the Dongguan site.
1-2. *pingdibo* bowls; 3. *gudibo* bowl; 4. *gudipen* basin; 5. *houdibo* bowl; 6. *houdipen* basin; 7-8. *qizuo* stands.

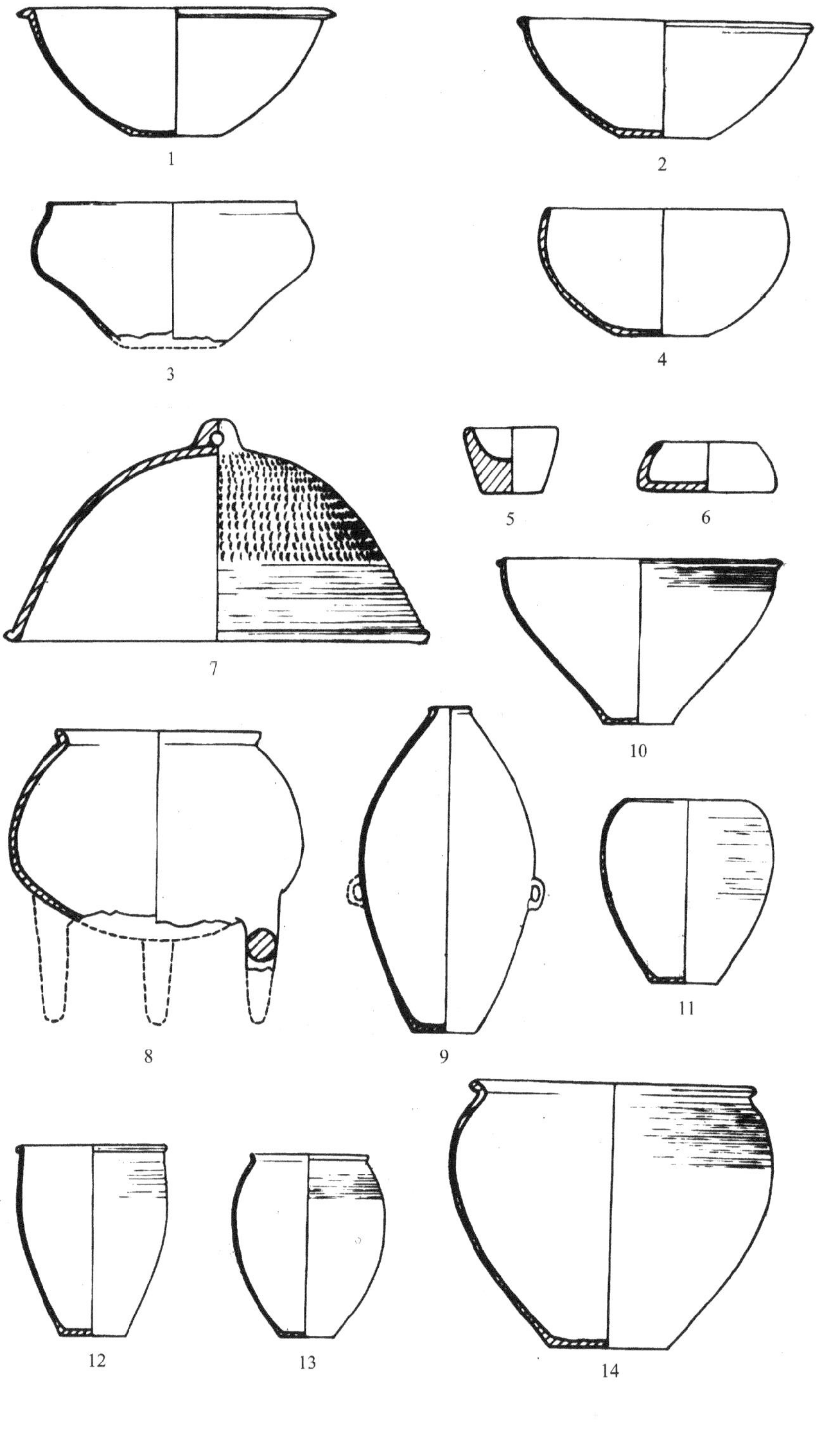

Figure 5-5b. Dongguan I ceramics from the Dongguan site.
1. *zheyanpen* basin; 2. *zhaiyanpen* basin; 3. *qufupen* basin; 4. *liankoubo* bowl; 5. *bei* cup; 6. *yu* dish; 7. *gai* lid; 8. *ding* tripod; 9. *ping* bottle; 10. *gang* jar; 11. *liankouguan* jar; 12-13. *chikouguan* jar; 14. *dakouguan* jar.

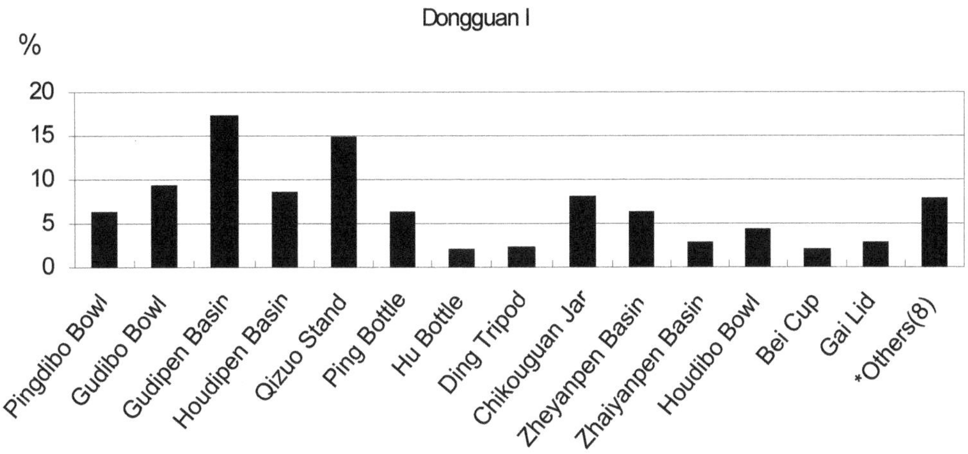

Figure 5-6. Distribution of ceramic classes in the Dongguan I assemblage.
*Others (8): Eight ceramic classes less than five specimens, including the *shuang'erhu* bottle, *gang* jar, *liankouguan* jar, *dakouguan* jar, *liankoubo* bowl, *qufupen* basin, *shenfupen* basin, and *yu* dish.

The richness and evenness of pottery classes in the Dongguan I assemblage will be further discussed in comparison with other assemblages, especially the Miaodigou II assemblage in the following chapter.

5.4. Indirect Evidence of Pottery Production

The studies in this section will follow the theoretical and methodological approaches outlined in Chapter Four. The indirect evidence includes nominal and metric variables of pottery. Both of them can be used to investigate standardisation/variability of pottery. The nominal data contain vessel forms (see last section), paste, colour, surface treatment and decoration, manufacturing techniques (mainly referring to building, shaping and finishing techniques), and so forth. Through the analyses on nominal and metric data, I will evaluate the degree of specialisation of pottery production, and examine the mode of production presented in section **5.2**.

5.4.1. Analyses of Nominal Data

It is necessary to firstly point out that there are no significant variations observed in nominal variables of pottery between the early and late phase, so I have put them together as one assemblage.

The pottery colours of Dongguan I include red, reddish-brown, grey, pale-brown, and black. These were the result of different firing atmosphere. The red and reddish-brown pottery was fired in oxidising atmosphere, the grey and pale-brown pottery was fired in reducing atmosphere, and the black pottery resulted from carbon-permeation technique (Li 2001:540). The red and reddish-brown pottery are more common than others, accounting for about 52% of the total, the grey and pale-brown pottery accounts for about 36%, and the black pottery accounts for about 12% (National Museum of Chinese History et al. 2001:47). In fact, there are also some subtle variations beyond these four colours. The variability and diversity of pottery colours reflect the unrestricted firing atmosphere or unskilled control to the firing atmosphere. Besides, importantly, there are no high correlations between ceramic classes and colours.

In contrast to the colour, the paste, manufacturing techniques, surface treatment and decoration had some correlation with the shape classes in this assemblage. However, these variables did not change over time during this period. There was no essential difference between the early and late phase.

The pottery paste of this assemblage includes two primary categories: silt paste and sandy paste. The majority of vessels are silt-paste pottery (accounting for 71% of the total), including all sorts of bowls, basins, bottles, cups, dishes, and stands. The sandy-paste vessels account for 29% of the total, including all sorts of jars and tripods (National Museum of Chinese History et al. 2001:47). It seems that all preparation and serving vessels are silt-paste, and all storage and cooking vessels are sandy-paste. Fabric is therefore related to shape or function and cannot be taken as an indicator of production system. The sandy paste also contains several secondary categories, such as coarse- and fine-tempered paste. Nevertheless, these secondary paste categories do not strictly correlate to secondary shape classes. For example, one type of jar may be made using several secondary categories of sandy paste.

All vessels of Dongguan I were built by hand, and mainly by coiling. Beating and scraping were the most common shaping techniques during this period. A notable phenomenon was the application of the "slow-wheel" shaping and finishing technique: the vessel was shaped and smoothed through slow rotation of the turntable, and rhythmic ridges and grooves spiralling around the vessel wall were left on the surface of the vessel (Li 2001:534-537). Marks of this kind of technique can only be seen on *houdipen* basins and *houdibo* bowls. They are invisible on other classes of pottery. Perhaps this technique was not applied to other ceramic types at all. It appears that there was some kind of special correlation between this technique and the two vessel forms.

The surface treatment and decoration are actually one component of finishing techniques. The vast majority of vessels in the assemblage are plain, including nearly all silt-paste vessels and some sandy vessels. Among silt-paste ceramics, all kinds of bowls, basins and bottles are highly burnished. For most *gudipen* basins, and a proportion of *gudibo* bowls and *pingdibo* bowls, there is a circle of red on the exterior surface of the rim (see Fig. 5-5a). This was the result of firing stacked vessels, rather than the remains of painted decoration. Besides, only some sandy jars were executed with cord-marking, string-marking, or bowstrings.

In summary, there are many correlations between the ceramic classes and the pottery paste, manufacturing techniques, and surface treatment and decoration in the Dongguan I assemblage. The diversity of all these nominal variables is relatively low as well. These phenomena are probably associated with technological constraints, entrenched custom, or social control (interference of the community or kin groups), rather than the results of highly specialised production (see the following analyses of metric data).

On the other hand, though the ceramics of Dongguan I are not complicated in the forms, they still exhibit good control and skilled manufacturing technology. For instance, several categories of silt-paste pottery, especially *pingdibo* bowls, *gudibo* bowls, and *gudipen* basins, have very thin vessel walls, and highly burnished surface. These vessels are elaborate and fragile. They may have been manufactured by the potters who were more skilled and experienced, but not by common members in households. The evidence in this aspect is thus counter to the productive mode of household production, but rather indicates some kind of specialised production.

5.4.2. Analyses of Metric Data

The direct purpose of these analyses is to investigate standardisation/variability of pottery dimensions, and then further evaluate the degree of specialisation of pottery production. In the case that direct evidence of pottery production is scarce or inadequate, the analysis of metric variables is perhaps the most effective way of assessing specialisation.

Standardisation can be effectively measured through descriptive statistics, including mean, standard deviation (SD), and the coefficient of variation (CV). The assessments will finally focus on more productive CV. In addition, I will use ANOVA (analysis of variance, F ratio) to examine the variability between two groups of vessels from the early and late phase.

I selected five ceramic types with relatively large samples for statistical analyses. They are the *pingdibo* bowl, *gudibo* bowl, *gudipen* basin, *houdipen* basin, and *qizuo* stand (see Fig. 5-5a). Although a few other ceramic classes, such as *chikouguan* jar, also have relatively large samples (see Fig. 5-6 and Appendix 2), most specimens are sherds rather than whole or recovered vessels. For some size variables, such as vessel height, we can only get a few figures, so these ceramic types have to be rejected from the analyses.

All measured samples of the five types of ceramics are the complete and recovered vessels. There should be a small number of fragmentary sherd samples for these ceramic classes according to the impression of two technological workers of the Gucheng archaeological team (personal communication), but I could not find them when I undertook the data collection in Yuanqu.

Because these five classes of vessels are all very simple in shape, I measured and recorded three size variables for them: orifice diameter, base diameter, and vessel height. In addition, I also recorded the ratio orifice diameter/height, which is regarded as the better variable for testing vessel uniformity (see Chapter Four).

In the following studies I will analyse the statistical values of the four variables from five ceramic classes, and compare variations between the two groups of vessels from the Early and Late phases (see Table 5-1).

As seen in Table 5-1, each ceramic type for the early and late phase has a wide range of CV values. For *pingdibo* bowls, *gudibo* bowls, and *gudipen* basins, all CV values in the late phase are higher than those corresponding values in the early phase. CV values for orifice diameter, base diameter, and height of *pingdibo* bowls in the early phase range from 5.54% to 12.86%; in the late phase from 19.31% to 22.76%. CV values for these three dimensions of *gudibo* bowls in the early phase range from 10.92% to 20.60%; in the late phase from 17.34% to 34.22%. CV values for the three dimensions of *gudipen* basins in the early phase range from 15.21% to 23.28%; in the late phase from 16.77% to 27.72%. On the contrary, for *houdipen* basins and *qizuo* stands, all CV values in the late phase are lower than those corresponding values in the early phase. CV values for the same three variables of *houdipen* basins in the early phase range from 19.61% to 25.31%, in the late phase from 8.50% to 22.58%. CV values for these dimensions of *qizuo* stands in the early phase range from 19.45% to 26.71%; in the late phase from 9.90% to 14.13%. These two converse patterns of variations indicate that CV values for different pottery types do not change along one way. The variations of CV values for these ceramics between the two phases may have not been the result of the change of the degree of specialisation or mode of production, but affected by other unknown factors. It is notable that a few groups of vessels, such as *pingdibo* bowls, *gudibo* bowls, and *houdipen* basins in the late phase are very small in the sample number (less than 10). The small number of samples is perhaps one factor resulting in the deviation of CV values, if this deviation exists. In general, the whole variable range of CV values for the three dimensions is from 5.54% to 34.22%. Apart from several extremely low and high values, most of them range from 10% to 27%. The spectrum of variation is very big, and the CV values are generally very high (Table 5-1).

Now let us review the variation of CV values for the ratio orifice diameter/height between the early and late phase. For three classes of vessels, *pingdibo* bowls,

Table 5-1. Sample means, standard deviations (SD), and coefficients of variation (CV) for the five classes of ceramics in the Dongguan I assemblage.

		Orifice diameter	Base diameter	Height	Orifice diameter / Height
Pingdibo Bowl					
Early	N=10				
	Mean (cm)	21.36	7.06	11.88	1.82
	SD	1.18	0.86	1.53	0.22
	CV	**5.54%**	**12.17%**	**12.86%**	**12.15%**
Late	N=6				
	Mean (cm)	24.48	8.52	13.32	1.86
	SD	4.98	1.64	3.03	0.24
	CV	**20.35%**	**19.31%**	**22.76%**	**12.85%**
Combined	N=16				
	Mean (cm)	22.53	7.61	12.42	1.83
	SD	3.40	1.37	2.23	0.22
	CV	**15.09%**	**18.00%**	**17.97%**	**12.02%**
Gudibo Bowl					
Early	N=15				
	Mean (cm)	17.81	6.81	6.69	2.73
	SD	1.95	1.35	1.38	0.38
	CV	**10.92%**	**19.85%**	**20.60%**	**14.03%**
Late	N=9				
	Mean (cm)	14.36	5.42	4.86	3.13
	SD	2.49	1.14	1.66	0.78
	CV	**17.34%**	**20.96%**	**34.22%**	**24.78%**
Combined	N=24				
	Mean (cm)	16.52	6.29	6.00	2.88
	SD	2.72	1.42	1.72	0.58
	CV	**16.45%**	**22.66%**	**28.57%**	**20.21%**
Gudipen Basin					
Early	N=26				
	Mean (cm)	29.38	9.22	13.13	2.28
	SD	5.01	1.40	3.06	0.26
	CV	**17.05%**	**15.21%**	**23.28%**	**11.27%**
Late	N=19				
	Mean (cm)	28.73	8.54	12.69	2.29
	SD	5.83	1.43	3.52	0.28
	CV	**20.30%**	**16.77%**	**27.72%**	**12.38%**
Combined	N=45				
	Mean (cm)	29.10	8.94	12.95	2.28
	SD	5.32	1.44	3.22	0.27
	CV	**18.27%**	**16.08%**	**24.87%**	**11.60%**
Houdipen Basin					
Early	N=15				
	Mean (cm)	28.30	14.58	12.47	2.31

		Orifice diameter	Base diameter	Height	Orifice diameter / Height
	SD	5.77	2.86	3.16	0.30
	CV	**20.38%**	**19.61%**	**25.31%**	**13.14%**
Late	N=7				
	Mean (cm)	29.71	13.64	12.21	2.47
	SD	5.67	1.16	2.76	0.27
	CV	**19.10%**	**8.50%**	**22.58%**	**10.92%**
Combined	N=22				
	Mean (cm)	28.75	14.28	12.39	2.36
	SD	5.64	2.46	2.97	0.30
	CV	**19.63%**	**17.20%**	**23.98%**	**12.54%**
Qizuo Stand					
Early	N=22				
	Mean(cm)	7.46	8.24	3.10	2.43
	SD	1.99	1.88	0.60	0.50
	CV	**26.71%**	**22.81%**	**19.45%**	**20.46%**
Late	n=16				
	Mean (cm)	8.06	8.64	3.15	2.57
	SD	1.14	1.16	0.31	0.39
	CV	**14.13%**	**13.48%**	**9.90%**	**15.01%**
Combined	N=38				
	Mean (cm)	7.72	8.41	3.12	2.49
	SD	1.69	1.61	0.50	0.45
	CV	**21.96%**	**19.16%**	**15.90%**	**18.23%**

Table 5-2. Analysis of variance (ANOVA) for the five classes of ceramics between the early and late phase in the Dongguan I assemblage.

		Orifice diameter	Base diameter	Height	Orifice diameter /Height
Pingdibo Bowl	F ratio	3.744	5.522	1.619	0.086
	P	0.073	0.034	0.224	0.774
Gudibo Bowl	F ratio	14.424	6.606	8.584	2.943
	P	0.001	0.017	0.008	0.100
Gudipen Basin	F ratio	0.396	2.434	0.188	0.008
	P	0.533	0.126	0.667	0.931
Houdipen Basin	F ratio	0.290	0.684	0.035	1.288
	P	0.596	0.418	0.854	0.270
Qizuo Stand	F ratio	1.162	0.568	1.109	0.958
	P	0.288	0.456	0.743	0.334

Note: Significant at $P<0.05$.

gudipen bowls, and *houdipen* bowls, the CV values for the ratio between the early and late phase are very similar to each other (see Table 5-1). The whole variable range of CV values for the ratio orifice diameter/height is from 10.92% to 24.78%. Apart from two extremely high values, most of them range from 10% to 15%. The spectrum of variation for the ratio is much smaller than that of the three single dimensions, and CV values of the ratio are commonly lower than those of the single dimensions. If the ratio value reflects the vessel variability or uniformity more effectively, the variability displayed by the ratio for these vessel forms is not as big as those displayed by the single dimensions, and the spectrum of variation of most ceramic classes between the early and late phase is not very big either. However, the CV values for the ratio are still generally high.

ANOVA (analysis of variance) is a more direct and productive method for investigating the variation among groups of vessels. F-ratios for each variable between the early and late ceramic vessels have been calculated, and evaluate the difference in variability at a given level (significant at $p<0.05$) (Table 5-2). Of these five ceramic classes, there are only significant variations for orifice diameter, base diameter, and height of *gudibo* bowls between the early and late phase ($p<0.05$). However, there is no significant difference for the ratio of orifice diameter-to-height of this vessel form ($p=0.100$). This pattern suggests that *gudibo* bowls do not vary significantly between the early and late phase when the size problems affecting vessel variability are eliminated by the ratio. In addition, for *pingdibo* bowls, only base diameter is significant at the 0.05 level for ANOVA. Except for these variables of the two vessel forms, there are no significant results for all other variables of the five ceramic types ($p>0.05$), so the null hypothesis of no difference between two groups of ceramics cannot be rejected (see Table 5-2). In summary, ANOVA for the five ceramic classes supports the conclusion that there is no statistically significant difference in variability between two groups of ceramics from the early and late phases in the Dongguan I period. This conclusion is basically consistent with the analytical results of CV values, especially on the ratio of orifice diameter-to-height, and also consistent with the analytical results of nominal data.

With this crucial conclusion, we can concentrate on the pooled data of the early and late phase, and assess the significance of CV values of every vessel form in the entire Dongguan I assemblage (see Table 5-1).

A remarkable phenomenon is that nearly all CV values of the combined data for each variable, including the ratio orifice diameter/height, are in the position between two values of the early and late phases. This situation also occurs in some ethno-archaeological case studies, in which all or most CV scores of pooled data are between the corresponding CV scores of two or more data sets (e.g., Longacre et al. 1988; Underhill 2003). The only exception in the Dongguan I assemblage is the CV value for the base diameter of pooled *gudibo* bowls, which is slightly higher than both CV values of the early and late phase. The whole variable range of CV values for orifice diameter, base diameter and height in the combined data is from 15.09% to 28.57%. Apart from an extremely high value, all other CV values in the three dimensions range from 15% to 25%. For the ratio of orifice diameter-to-height in the combined data, the variable range of CV values is from 11.60% to 20.21%. Three of them range merely from 11.60% to 12.54%, and are very approximate to one another (see Table 5-1). Obviously, the spectrum of variation of CV values for the combined data is much smaller than that for the early and late phases. Nonetheless, most CV values for the combined data are still relatively high, and indicate relatively low uniformity and standardisation.

Since the number of samples for every ceramic class in the combined data increases substantially, ranging from 16 to 45 (four of five ceramic types are more than 20 in number), the statistical values are probably more exact and reliable than those for just the early or the late phase.

The degree of standardisation reflected by CV values, and the significance of this degree for mode of production can be efficiently assessed only in comparison with other archaeological or ethnological cases. However, it is not easy to find a good comparative example with the similar condition as the Dongguan I assemblage. Especially for ethnological cases, they usually address pottery samples from individual potters, or at most a collection of products from some potters in a short time (ranging from several days to several years). There are few comparisons between the ethnographic examples and archaeological cases with a long-time span, so the archaeological case studies are preferable for comparison. Though the following case study is not ideal for comparison, it is still helpful to illuminate the Dongguan I data.

Allen (1992) studied the pottery production by Iroquois, craftspersons, members of a "tribal" society. The ceramic sample consisted of material from six sites within one tribal area that was occupied for a period of about 80 years (Allen 1992:148). The direct evidence of pottery production here is scanty, but the relatively high coefficient of variance for pottery metric data indicates that household production is the most likely mode of production for Iroquois. The measurement and analyses focused on the collared cooking-pots. Because the ceramics examined in this case were fragmentary, collar height and rim thickness were measured as primary indicators. The author calculated CV values for thickness, collar height and the ratio collar height/thickness for the vessels from each site. CV values for the three size categories of vessels (small, medium and large) range from 12.4% to 37.0% (except an extremely high value 51.9%), and most of them range from 20% to 35% (Allen 1992:152). Most CV values in this case are higher than those in the early and late phases, as well as the combined data in Dongguan I assemblage. We may expect slightly higher CV values for Iroquois vessels if the time span of the ceramic assemblages increased to around 200 or 400 years like the Dongguan case. Because CV values for the Dongguan I assemblage are commonly lower than those for Iroquois ceramics, household production most likely existing in the Iroquois society may have been precluded from the Dongguan I society. The pottery production in the Dongguan I society was probably conducted at a more advanced organisation. On the other hand, the degree of standardisation and specialisation indicated by CV values for the Dongguan I assemblage is obviously lower than that from the nucleated workshops. This will be detailed in Chapter Seven. Therefore, it may be reasonable to infer that the pottery production in the Dongguan I society was a household industry or took place in individual workshops. It is regretful that I could not find comparable statistical data for these two kinds of productive modes in archaeological case studies.

In brief, we can at least draw two important conclusions through the analyses above. First, there is no significant variability for ceramic dimensions between the two groups of pottery from the early and late phases in the Dongguan I period. Namely, the degree of ceramic standardisation and specialisation of production remained at a similar level between the two phases, and the mode of production did not change essentially over time during this period. Second, the relatively high CV values for both individual phase and grouped data seem to indicate low standardisation and specialisation. The mode of production reflected by these statistical values appears to fit with household industry or individual workshops rather than others.

Finally, let us review the variance of CV values for different metric variables of the five ceramic classes in the Dongguan I assemblage (see Table 5-1). For orifice diameter, most CV values range from 10.92% to 21.96%, apart from two extremely low and high values (5.54% and 26.71%). For base diameter, CV values mainly range from 12.17% to 22.81% (except an extremely low value 8.50%). For height, most CV values range from 12.86% to 28.57%, apart from two extremely low and high values (9.90% and 34.22%). For the ratio of orifice diameter-to-height, CV values mainly range from 10.92% to 20.46% (except an extremely high value 24.78). Comparatively, CV values for orifice diameter and base diameter are approximate to each other, and most of them are less than 20% (10%-20%), whereas most CV values for height are more than 20% (20%-28%). In contrast, CV values for the ratio orifice diameter/height are relatively low, and are the most concentrated, as most of them concentrate on 10% -15%. Therefore, height is the most variable variant for most vessel forms in this assemblage, and the ratio is the most conservative variable. Once again, the latter is confirmed that it can better reflect the variation of various ceramic classes by eliminating the size problem affecting vessel uniformity. In a word, these analytical results will be useful for comparison with the following studies.

5.5. Summary

The preceding analyses evaluate the mode of production in the Dongguan I society through direct and indirect evidence of pottery production. Here I further summarise the properties of the mode of production from context, concentration, scale, and intensity of production.

In the light of direct evidence of pottery production, especially the presence of pottery kilns, I could firstly infer that Dongguan I ceramics were mainly manufactured at the Dongguan site. From the distribution pattern of these kilns and other evidence, including the indirect evidence of pottery production, and in comparison with several archaeological and ethno-archaeological cases, I conclude that household industry or individual workshop industry was the possible mode of production for the Dongguan I settlement. Because there were no definite workshops discovered in Dongguan and a few other contemporary sites uncovered on a large scale in the Wei River valley, pottery may have been mainly produced in domestic context. If that is true, household industry may have been the major organisation of production for this period.

In terms of the number, size and distribution pattern of kilns at Dongguan and the other contemporary sites, the pottery production in the Early Yangshao societies was probably small-scaled and dispersed, at both communal and regional levels. It is likely that a small number of households or communal members produced pottery for their own community and some other non-pottery-producing communities.

Regarding intensity of production, it is almost impossible to find direct evidence in archaeological cases, but we can infer from indirect data. In the Dongguan assemblage, relatively high CV values for pottery dimensions seem to indicate the lower degree of standardisation and specialisation of production. The patterns of these statistical data possibly correspond the small-scale, dispersed and part-time specialised production. The pottery products may have been manufactured by part-time or seasonal specialists or potters. Certainly, the significance of these statistical data can only be showed more clearly in the following chapters by comparison. Actually, according to the definition of household industry and individual workshops given by Peacock (Peacock 1982:8-9), potting is mainly a part-time activity under these two modes of production. These two kinds of productive modes are similar in many aspects, and can also be together named as "household specialized production" (Stark 1992:184).

With regard to the context of production and identity of producers, one possibility is that potters were completely independent and produced pottery for exchange as the complement of their subsistence or as the primary source of their livelihood. The other possibility is that producers were subsidiary to their communities or kin groups on the basis of divisions of labour, and the production and distribution of pottery was thus organised and managed by communities or different kin groups. The latter can be supported by the linkage between the kilns and different house clusters (kin groups) at Beishouling and Jiangzhai, the Early Yangshao settlements in the Wei River valley.

Finally, another important conclusion reflected in indirect evidence of pottery production (including nominal and metric variables) is that there is no essential difference in variability between the two groups of ceramics from the early and late phases during the Dongguan I period. The mode of production may have sustained continuity during this period.

Chapter Six

Pottery Production in a Chiefdom-level Society: A Case Study of the Miaodigou II Period at the Dongguan Site

6.1. Introduction

During the Miaodigou II period (ca. 2900-2500 B.C.), a complex society had developed in the Yuanqu Basin. Although the evidence for the specific social form is not very adequate, the chiefdom-level society may be a reasonable proposition for this regional polity (see Chapter Three). This chapter will discuss the mode of pottery production in such a society, mainly based on the pottery data of the Miaodigou II period at the Dongguan site.

The Dongguan site was the largest settlement (20 ha in area) and the central place during the Miaodigou II period in the Yuanqu Basin. The excavation report of this site subdivides the remains of Miaodigou II into three chronological phases (the early, middle and late phase) in the light of the variance of assemblage composition of ceramic classes and vessel shape (National Museum of Chinese History et al. 2001). Each phase spanned more than 130 years. I will investigate the differences in pottery production between the phases.

As in the previous chapter, I will reconstruct the mode of production of this period through direct and indirect evidence, and assess the changes between this period and the Dongguan I period. Besides Dongguan, the direct evidence of pottery production will involve Ningjiapo, another contemporary site in the Basin, which is very helpful for illuminating the production organisation at Dongguan.

6.2. Direct Evidence of Pottery Production

Of five Miaodigou II sites excavated in the Yuanqu Basin, four have direct evidence of pottery production. There is a pottery anvil and a pottery mould (used for making legs of *jia* tripods) from Fengcun (Shanxi Team 1987:40, 42) (Fig. 6-1). A pottery kiln was uncovered at Bancun (Railey 1999:438). Two well-preserved kilns were found at Ningjiapo (Shanxi Institute of Archaeology 1998). For the Dongguan site, the direct evidence of pottery production is relatively abundant. There are five pottery kilns in the early phase, one anvil in the middle phase, and two anvils in the late phase (National Museum of Chinese History et al. 2001) (see Fig. 6-1). No direct evidence of pottery production was recovered at Xiaozhao. This may be due to the poor survival of the Miaodigou II remains and the small excavated area at this site (Shanxi Team 1998, 2001), so it is still possible that pottery was produced in this settlement. These finds suggest that most, if not all, communities were engaged in pottery production during the Miaodigou II period in the Yuanqu Basin, and that pottery production was dispersed at the regional level, rather than concentrated in a single community.

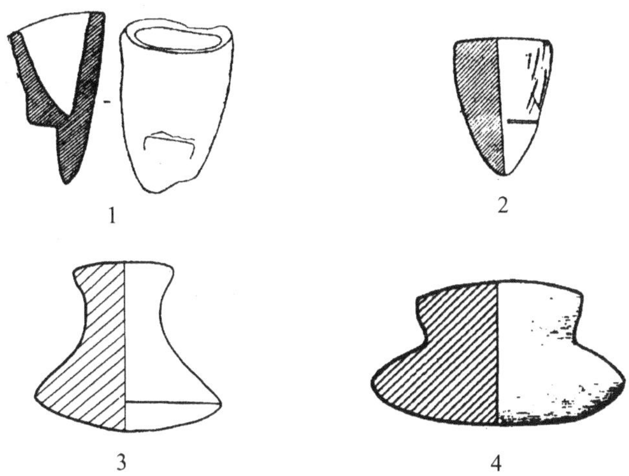

Figure 6-1. Pottery-manufacturing tools of the Miaodigou II period from the Fengcun and Dongguan site.
1. anvil (Fengcun); 2. mould (Fengcun); 3-4. anvils (Dongguan).

Before discussing the Dongguan case, I intend to firstly examine the direct evidence of pottery production at the Ningjiapo site, which provides a better example for understanding pottery manufacture during this period.

An important discovery at Ningjiapo is that a workshop may have appeared during the Miaodigou II period. Two well preserved pottery kilns were uncovered at Ningjiapo, and fortunately the data on both have been reported in detail (Shanxi Institute of Archaeology 1998). The Ningjiapo site is about three kilometres northwest of the Dongguan site, at the southern side of the Boqing River (see Fig.3-16). This site measures around six hectares. During the Miaodigou II period, a wide and deep ditch oriented west-east crossed the site. The northern section of the site between the ditch and the riverbank was the major area for feature distribution, and was possibly the central area of the settlement. The ditch measures 9-10 m in width, and 2.6-4.2 m in depth. It slants down from the west to the east on the bottom. It has been revealed for over 15 m, but its total length is not yet clear. Two elaborate and very well preserved kilns (Y501 and Y502) were recovered on the north wall of the ditch (Fig. 6-2). The distance between the central points

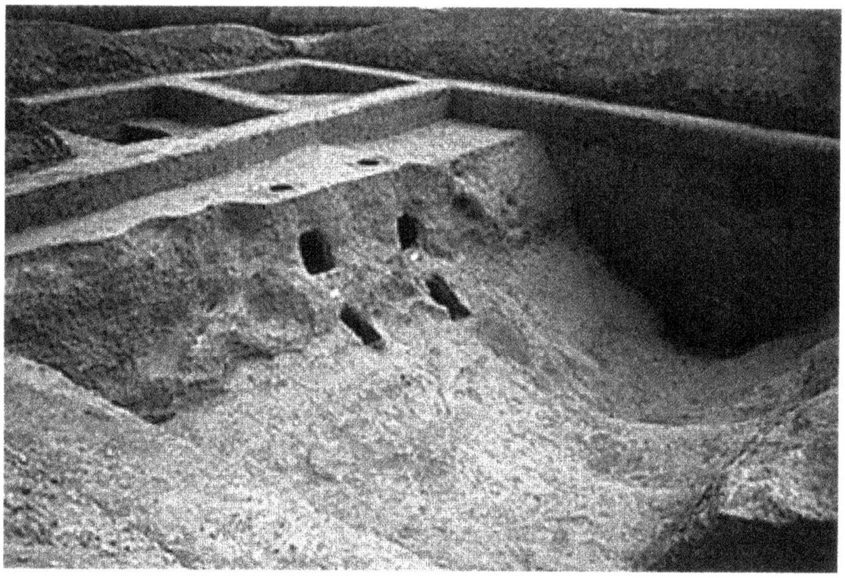

Figure 6-2. Location and surface of the kilns Y501 and Y502 at Ningjiapo (after Shanxi Institute of Archaeology 1998:Colour Page 3:1).

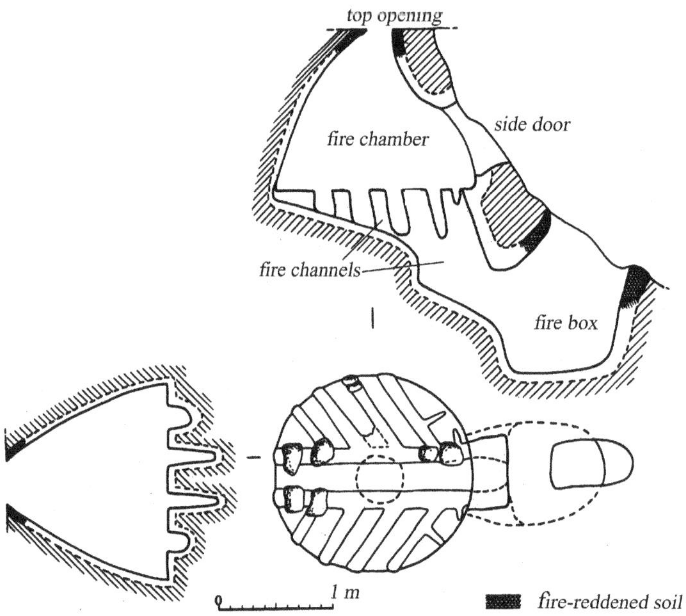

Figure 6-3. Plan and cross-sections of the kiln Y501 at Ningjiapo (after Shanxi Institute of Archaeology 1998: fig. 2).

of two kilns is 2.3 m, and the closest distance between their edges is merely 0.5 m, so they are very close together each other, and most likely belonged to one production unit.

The structures of the two kilns are almost the same and very sophisticated. They both consist of a fire box (the lower portion of the kiln), fire channels, a firing chamber (the upper portion of the kiln), a side door to the firing chamber, and a top opening for the firing chamber which is built on the ground surface above the ditch (Fig. 6-2 and 6-3). The fire channels are composed of two major and several tributary fire courses. The top opening of the firing chamber could be used for altering firing temperature and for water permeation into the kiln room. All these kiln components were formed by digging into the north wall of the ditch, so only the opening of the fire box, the side door of the firing chamber, and the top opening of the firing chamber are exposed on the surface (see Fig. 6-2). The two kilns are also similar in size. Y501 is 1.7 m and Y502 is 1.9 m in bottom diameter of the bell-shaped firing chamber. The elaboration and complication of the two kilns seem to indicate some degree of specialised production rather than casual, occasional production.

There is a platform functioning as a possible working area in front of the two kilns at the bottom of the ditch. This platform measures about 30 m^2 in area, and slightly slants down to the east. Some firing ash was left on its surface. A path runs from the working area, through five steps along the north wall of the ditch beside Y501 (the west one), to the ground surface above the ditch (see Fig. 6-2).

At the south of the working platform a large pit was dug into the south wall of the ditch. This pit measures seven meters in north-south width, and 2.6 m in maximum depth from its bottom to the ground surface. The west-east length exposed in the excavated block is 4.2 m, but the total length is not clear because the rest of the pit is outside the excavated area. The cross section of the pit shows that below yellow loess is a deposit of red soil, which is fine and sticky, and very appropriate to function as clay for potting. This kind of red soil is still used for potting by many modern potters in some loess areas in the Middle Yellow River valley (I once witnessed this in a field survey in early 2001 in Gongyi, Henan). In addition, of particular importance is that there was a layer of clay measuring 0.5 m in thickness at the bottom of the pit. The clay was very fine and pure without any impurity. Therefore, the excavation report contends that this was a special place for acquiring and preparing clay. Correspondingly, pottery fabrication may have been carried out nearby.

As discussed in the excavation report, all these foregoing features constitute a complete picture for manufacturing and firing pottery. Accordingly, it should be reasonable to argue that these features formed a workshop. This workshop perhaps also involved some other undiscovered or destroyed facilities, such as the buildings for making and storing pottery. Due to the limitations of the excavated area, we cannot be sure if there were other workshops or how many workshops like this existed during the Miaodigo II period in the Ningjiapo settlement.

As mentioned above, the central area of the Ningjiapo settlement was probably situated between this southern ditch and northern bluffs (the riverbank), so it is possible that the ditch functioned as the fortification and drainage system preventing flood from the southern highland to the main residential zones. Hence the construction of the two kilns just took advantage of the bluff of the ditch, and this workshop was also located at the edge of the village. It is likely that pottery production could not be performed during the rainy seasons.

The Ningjiapo case enables us to better understand the pottery production of Dongguan for this period.

The five kilns of the early phase at Dongguan are located in a small excavation area in Area III (Fig. 6-4). At the western section of the excavated area there are no kilns, so I infer that no kilns exist to the immediate west beyond the excavated area. The north and south sides of the excavated area are close to bluffs, so it is also unlikely that there are kilns in these areas, or the kilns once existing there have been destroyed with the erosion of the terrace. The uncovered five kilns are all distributed in the eastern section of the excavated area. As there is a relatively large area (around 400 m^2) unexcavated between the excavated area and the eastern bluff of the terrace, it is possible that a few other kilns existed in this area, at the east of the excavated area. In general, here could be a major area for pottery firing. Similar to Dongguan I, this pottery-firing area was also close to the eastern bluffs of the terrace, at the eastern edge of the village, and probably outside the residential areas.

Within this area, however, the five kilns are not close together, but are a similar distance from each other, ranging from 6.5 to 8.5 m. The kilns are all very small, ranging from 0.5 to 1 m in diameter at the base of the kiln room. In such a small kiln, only 1-2 larger vessels or 4-8 small vessels could be contained and fired at the one time (Institute of Archaeology 1963; Xi'an Banpo Museum et al. 1988). On the other hand, apart from Y1, each of the other four kilns is linked to a large pit that functioned as the working area (see Fig. 6-4). They were built immediately beside the pits, and their fire boxes were dug into the pit walls and connected with the pits. These pits are all irregular in shape, measuring around five meters or more in maximum diameter, and around one meter or more in depth. Every pit yielded some ash, fire-reddened soil, and small sherds related to firing.

For example, IIIY4 is a well-preserved kiln (National Museum of Chinese History et al. 2001:181). The firing chamber is slightly bell-shaped, measuring 0.54-0.5 m in diameter at the base and 0.55 m in height. There is a small opening on the top of the chamber, measuring 0.16 m in maximum diameter. Beneath the firing chamber is the fire box (Fig. 6-5). The opening of the fire box is at the base of the kiln, and was formed on the eastern wall of IIIH22 (Fig. 6-6). IIIH22 is an irregular large pit measuring 6.6 m in maximum diameter and 2.08 m in maximum depth. This pit is very complicated in structure, consisting of a doorway, three steps, a slope, a pit for firing residue linked with the fire box, a large and two small earth hollows, and so forth (see Fig. 6-6). The large hollow measures 2.1 m in width of the opening, 1.3 m in height, and 1.7 m in horizontal depth. It could be used to store fired and unfired pottery products. In addition, these three hollows may have been formed by procuring clay for potting from the pit walls where there is the similar deposit of red soil as in Ningjiapo. At the south-western corner of the pit is a shallow pit with the flat bottom, measuring 2.24 m in length, 1.52 m in maximum width, and 0.12-0.2 m in depth, which could have been used for clay preparation as in Ningjiapo. Generally speaking, the structure of this kiln and its working area is very similar to the Ningjiapo example. Therefore the excavation report infers that IIIH22 was also the location for pottery manufacture, and the clay for pottery making was procured directly from the pit (National Museum of Chinese History et al. 2001:181). In that case, this kiln and pit, as well as other kilns and linked pits may have constituted workshops at Dongguan.

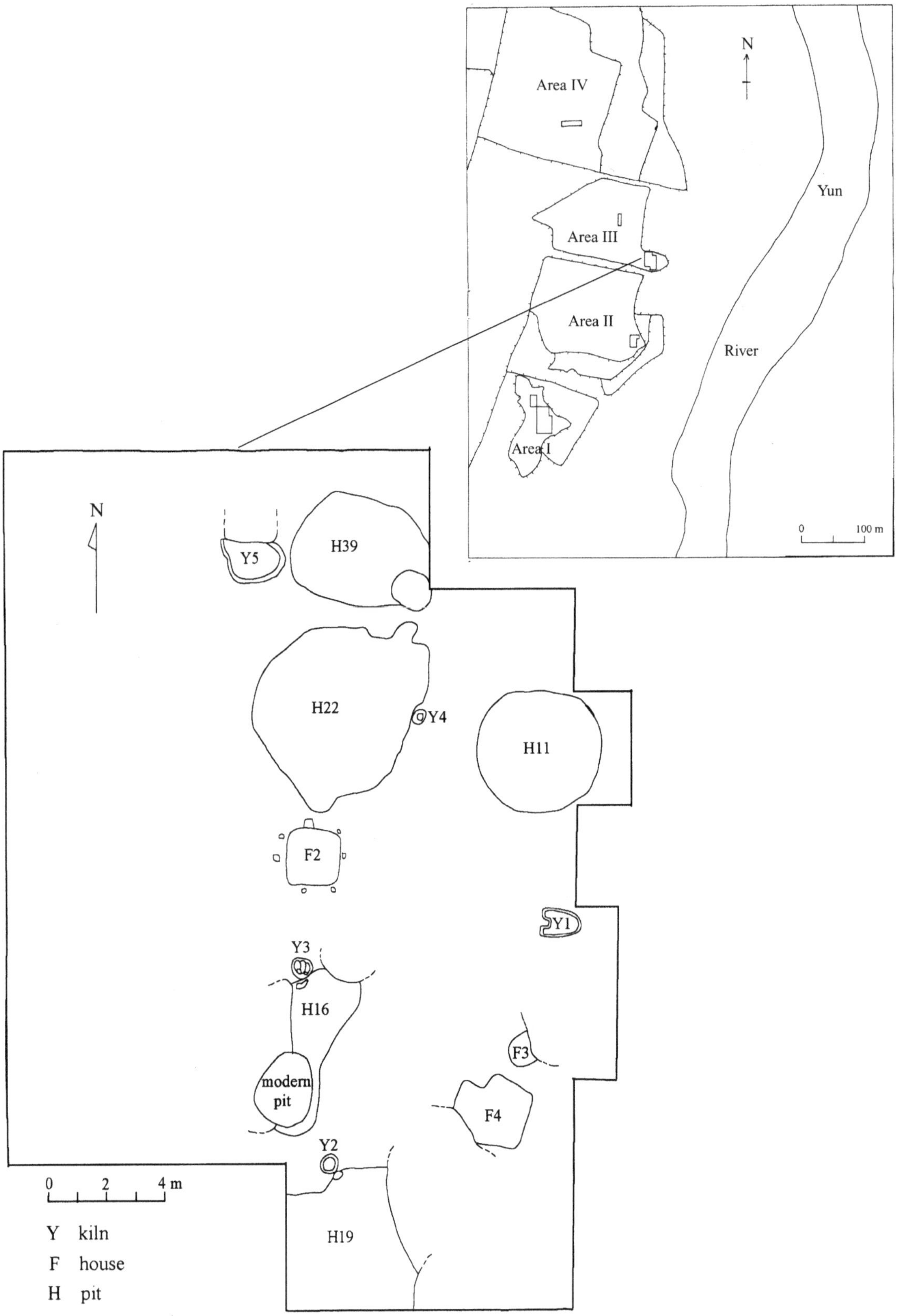

Figure 6-4. Distribution of Miaodigou II kilns and relevant features at Area III of Dongguan (redrawn from National Museum of Chinese History et al. 2001:fig.9).

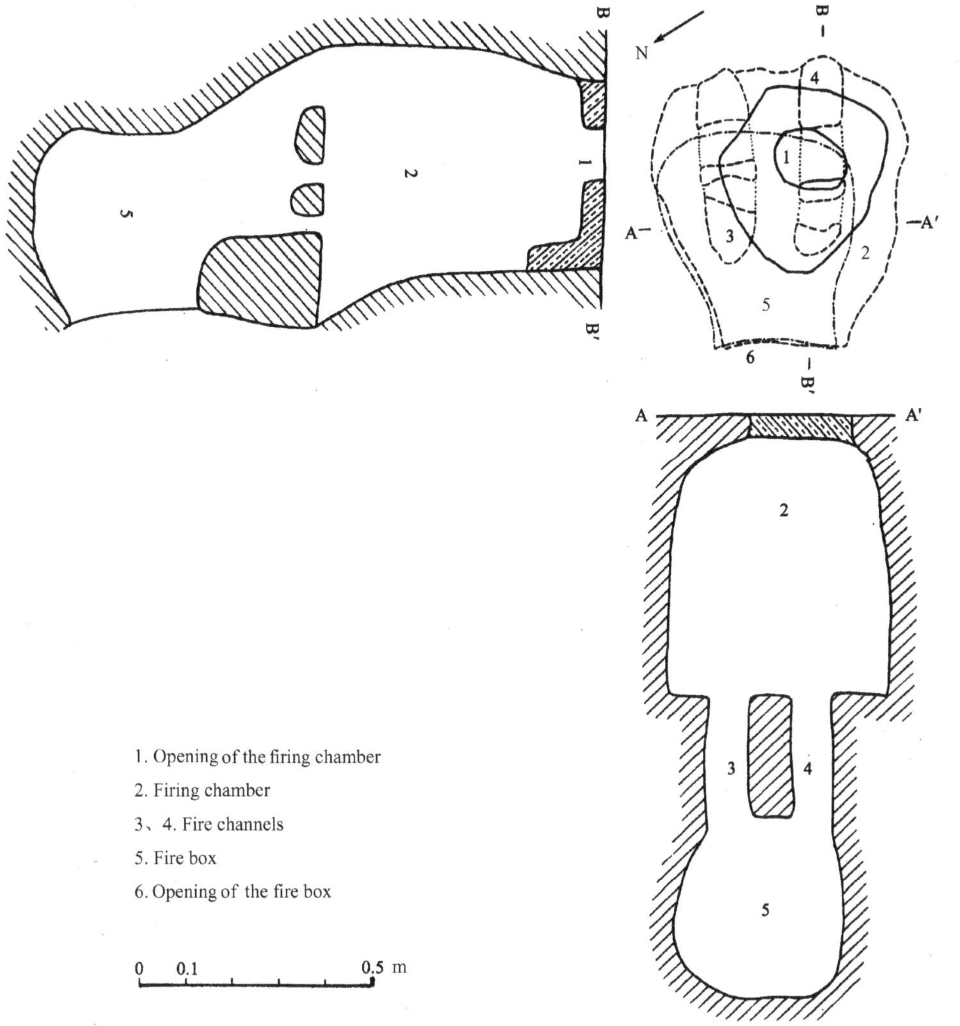

Figure 6-5. Plan and cross-sections of the kiln IIIY4 at Dongguan (after National Museum of Chinese History et al. 2001:fig.154).

In addition, there were also some special storage pits near these kilns. For example, IIIH11 provides sufficient evidence in this aspect (National Museum of Chinese History et al. 2001:175-176). This pit is approximately circular in shape, measuring around 4.5 m in diameter, and 2.12 m in depth. There is a platform measuring 0.25-0.3 m in height at the eastern part of the pit. A long earth ladder runs along the north wall from the ground surface to the bottom of the pit (Fig. 6-7). This pit yielded a large number of sherds, from which 52 vessels were recovered, including a range of vessel forms (nearly 20). This suggests that potters did not specialise in making a small number of pottery types. The excavation report argues that most of these vessels were destroyed at their original position in the pit. This situation and the structure of the pit suggest that IIIH11 may have been a storage facility associated with nearby kilns (such as Y4 or Y1, see Fig. 6-4).

Three single semi-subterranean houses dating to the early phase were discovered near the kilns in this excavated area (see Fig. 6-4). The three houses are all very small, ranging from 2.3-5.4 m^2 in area, and there were few facilities and wares for domestic use found in them, so it is unlikely that these structures were household residences. IIIF3 (2.3 m^2 in area) has a small stove pit at the north-eastern section of its bottom, so the excavation report concludes that this was a separate kitchen (National Museum of Chinese History et al. 2001:169). This kitchen could have been used to cook food for potters during the production seasons. The function of IIIF2 (4.4 m^2 in area) and IIIF4 (5.4 m^2 in area) is not very clear. It is likely that they functioned as storage warehouses for pottery or even workshop structures for pottery manufacture.

Based on our observation in early 2001, the modern potting workshops in Gongyi, Henan are very similar to the archaeological examples at Dongguan. All workshops are located in places near clay sources (red soil deposits). In one case, the clay is available in the courtyard of the workshop. Each workshop contains all the facilities for

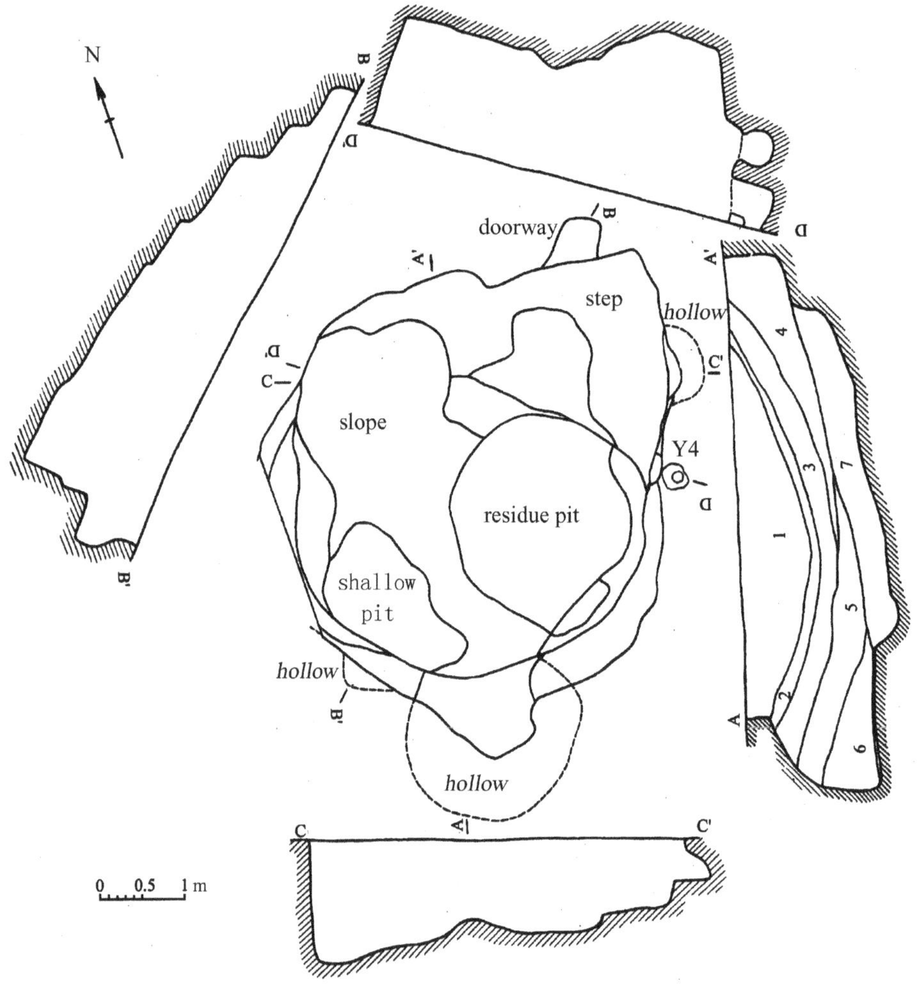

Figure 6-6. Plan and cross-sections of the pit IIIH22 at Dongguan (after National Museum of Chinese History et al. 2001:fig.155)

pottery production, including the workshop structure, storage warehouse, kilns, and so on. Some workshops also have kitchens.

The workshops in the Yuanqu Basin are not an isolated example for the late Neolithic in China. In other regions of the Yellow River valley, some sites dating to the Miaodigou II and Longshan periods not only had similar potting facilities as Dongguan and Ningjiapo, but also provided more complete archaeological evidence on pottery workshops. For instance, at Laohushan in Liangcheng, Inner Mongolia (Inner Mongolia Institute of Cultural Relics 2000:342-358), and Baidaogouping in Lanzhou, Gansu Province (Gansu Committee of Cultural Relics Management 1957; Yan 1989b), both sites had the specific area for specialised pottery production, where there were kilns and relevant large pits (working area), pits for preparing clay, prepared clay and clay strips for coiling pottery, platforms for making pottery, and unfired ceramic vessels. Likewise, there were no pottery-making tools found there, which could often be easily taken away from workshops.

Several houses dating to the early phase, and some houses dating to the middle and late phase were found at Area I and Area II at the Dongguan site (see Fig. 3-6 for the locations of the two areas). A few of them are relatively large and yielded many items associated with domestic activities and agricultural production (such as IF8 in the early phase, IF3, IF4 and IF6 in the middle phase). Such houses were especially concentrated in Area I, where a residential section seemed to exist. However, no direct evidence relevant to pottery manufacture, such as tools and clay, were discovered inside these houses, and no productive facilities such as kilns were uncovered near these houses. Therefore, it is unlikely that pottery production was conducted in domestic contexts. In that case, at least household production, and even household industry could be ruled out from the mode of production in Dongguan during the Miaodigou II period. However, due to the limitation of excavated area, we cannot negate the possibility that a few households were engaged in pottery manufacture in the domestic context, so the household industry cannot be completely rejected.

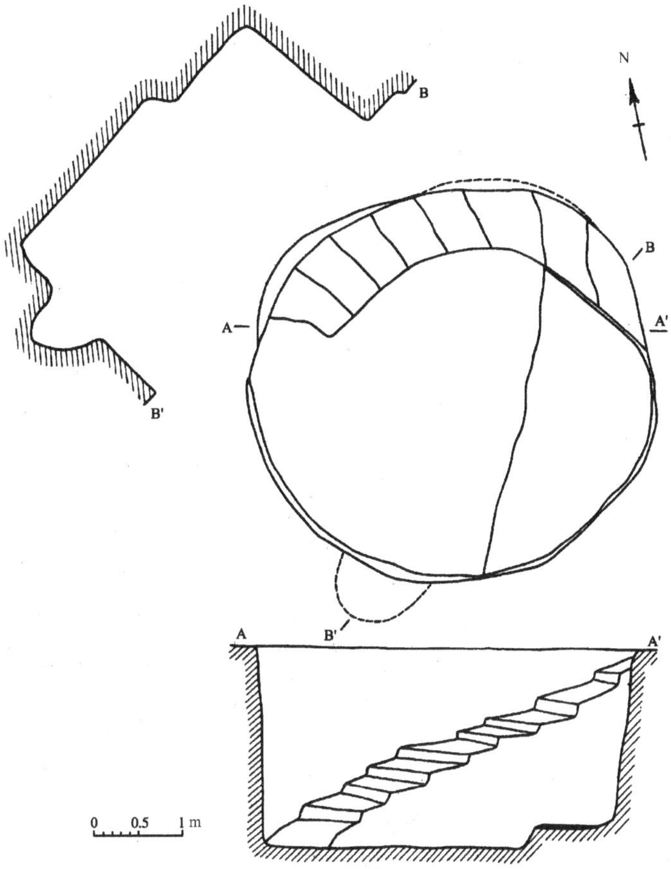

Figure 6-7. Plan and cross-sections of the pit IIIH11 at Dongguan (after National Museum of Chinese History et al. 2001:fig.148)

The workshop consisting of the two kilns and other relevant facilities at Ningjiapo formed a single production unit. Perhaps there were also other such workshops or units inside or outside the ditch, but they were bound to keep a relatively long distance from this workshop because they were not visible within the excavated area. We may refer to these workshops as "individual workshops". At Dongguan, the five kilns (or more) were collectively distributed in one area, but in the meantime kept a certain similar distance to each other. These kilns and their working areas may have constituted separate and independent production units as that of Ningjiapo, and also belonged to individual workshops. If so, the major mode of production at Dongguan and Ningjiapo during the Miaodigou II period can be identified as "individual workshop industry" (see Chapter Four).

On the other hand, the patterns of kiln distribution at Dongguan (some kilns, namely some workshops clustered together) are apt to remind us of "nucleated workshops" (see Chapter Four). Nevertheless, this possibility can be rejected by the indirect evidence of pottery production—standardisation index of metric data— which will be discussed later.

Like the situation in the Dongguan I period, the social context of pottery production is uncertain in this period. During the Miaodigou II period, social stratification had commonly developed in many regions (Yan 1999), and status and roles of families (including nucleated and extended families) in societies had become increasingly significant (Zhao 2001:107,127). The chiefdom-level society in the Yuanqu Basin may have been in the same case. On the other hand, there is no evidence of prestige pottery products during this period in the Basin (see the following analyses), and there was no particular linkage discovered between the kilns/workshops and specific house groups. Accordingly, it is very possible that pottery production was conducted by independent family members or work groups including some non-family members, but not manipulated or managed by either the elite, or large kin groups as probably exhibited in some Early Yangshao societies (see Chapter Five).

If every workshop in Yuanqu was independent, the scale of production for these individual workshops could be small. Among them, the workshop recovered in Ningjiapo was possibly the largest with two relatively big kilns. At the community-level pottery production, as the central settlement in the Basin, Dongguan may have owned the largest scale of production with some small individual workshops working together.

No direct evidence, such as a large quantity of refuses or kiln wasters, indicates a high intensity of production at Dongguan and Ningjiapo. The pottery production during this period was probably still part-time or seasonal, rather than full-time or year-round. This will be further confirmed by the following study on metric data of pottery.

6.3. Pottery Classification and the Assemblage Composition

Though many major ceramic types remained stable through time, there is some difference in the assemblage composition of ceramic classes between the three phases of the Miaodigou II period, so I discuss pottery classification and constitution in the three phases separately, and assess the variation in richness and evenness of ceramic types between them.

There is a total of 1018 specimens recorded for the whole Miaodigou II assemblage at the Dongguan site. Of these, 375 specimens are from the early phase, 244 from the middle phase, and 399 from the late phase. In terms of shape and size traits, I identify 33 ceramic types for the early phase, 27 for the middle phase, and 31 for the late phase (Appendix 3 and Fig. 6-8 ~ Fig. 6-11). The richness of ceramic classes for the three phases is approximately the same within each phase. Accordingly, the assemblage constitution of ceramic types did not display a remarkable variance of diversity over time during the Miaodigou II period. On the other hand, compared with the number of pottery categories (22) in the Dongguan I period, the richness of form classes obviously increased during this period, especially in the Early Miaodigou II phase, and this probably reflected the increase of social requirements for pottery use in a complex society.

The distribution of ceramic classes in the assemblages of the three phases is also similar to each other (Fig. 6-11). For instance, the top five pottery types in the early phase account for 41.6% of the total, ranging from 6.93% to 10.93% (including the *penxingding* tripod, *xiaokouguan* jar, *xiaoguan* jar, *daguan* jar, and *aoxinpen* basin). The top five pottery types in the middle phase account for 49.6% of the total, ranging from 5.74% to 15.98% (including the *guanxingding* tripod, *xiaoguan* jar, *daguan* jar, *bo* bowl, and *jia* tripod). The top five pottery types in the late phase account for 42.11% of the total, ranging from 6.77% to 12.53% (including the *guanxingding* tripod, *daguan* jar, *bo* bowl, *fuzao* cook stove, and *houdiwan* bowl). Most pottery classes for the three phases are less than 5% of the total (see Fig. 6-11 and Appendix 3). In general, the distribution of ceramic categories in the assemblages of the three phases is all decentralised rather than concentrated, and is high in evenness. The evenness of ceramic classes in the Miaodigou II assemblage is even higher than that in the Dongguan I assemblage (see Chapter Five). This pattern indicates that pottery production was not specially concentrated on several categories of ceramic vessels during the Miaodigou II period.

The richness and evenness of ceramic classes for the three phases indicate high diversity for the assemblage composition during the Miaodigou II period. Besides the influence of social function and demands, this could be attributed to a relatively low level of specialisation. This conforms to the wide range of pottery forms produced by each kiln as discussed above (see **section 6.2**, the case of IIIH11).

Judging from diverse aspects, including vessel forms, appearance, surface treatment and decoration, and context in the site, there were no prestige goods signalling elite ranks and status in the Miaodigou II assemblage of the Dongguan site. Therefore, the emergence of complex societies was not necessarily accompanied by the attached organisation of production manufacturing prestige goods (Stein 1996).

6.4. Indirect Evidence of Pottery Production

As in the previous chapter, the analyses of indirect evidence of pottery production involve nominal and metric data, by which standardisation/variability of pottery products can be investigated, and the degree of specialisation or the mode of production can be further evaluated.

I measured and recorded nominal and metric data of eight categories of ceramic vessels for statistical analyses, including the *ding* tripod, *xiaokouguan* jar, *xiaoguan* jar, *daguan* jar, *bo* bowl, *kecaopen* basin, *shuangbanpen* basin, and *shuangbanbo* bowl (see Fig. 6-10~6-12). However, some of them, especially the last three, have to be often rejected from different analyses due to the small sample sizes.

6.4.1. Analyses of Nominal Data

The analyses of nominal data comprise pottery colour, texture of paste, surface treatment and decoration, and manufacturing techniques, as mentioned in the previous two chapters.

The pottery colour for the Miaodigou II assemblage at the Dongguan site is a variable that cannot be exactly analysed quantitatively, because different colours often appear on a single vessel, and it is sometimes difficult to identify the original colour of a vessel, especially from a sherd. This point has been discussed in detail in Chapter Four. On the other hand, the proportions of various colours in the assemblages and their variations between the three phases can be investigated through calculating the proportions of sherds with different colours in the assemblages. According to the excavation report, although the assemblage composition of pottery colours changed a little over time, grey and pale-brown composed the majority of pottery colours for the three phases. In addition, there were a small number of red and reddish-brown, and a few black vessels (Table 6-1) (National Museum of Chinese History et al. 2001:184,247,292). There were no significant correlations between colours and vessel forms. Converse to the Dongguan I assemblage, most ceramics (grey and pale-brown pottery) in the Miaodigou II assemblage were fired in reducing atmosphere rather than oxidising atmosphere (Li 2001:553).

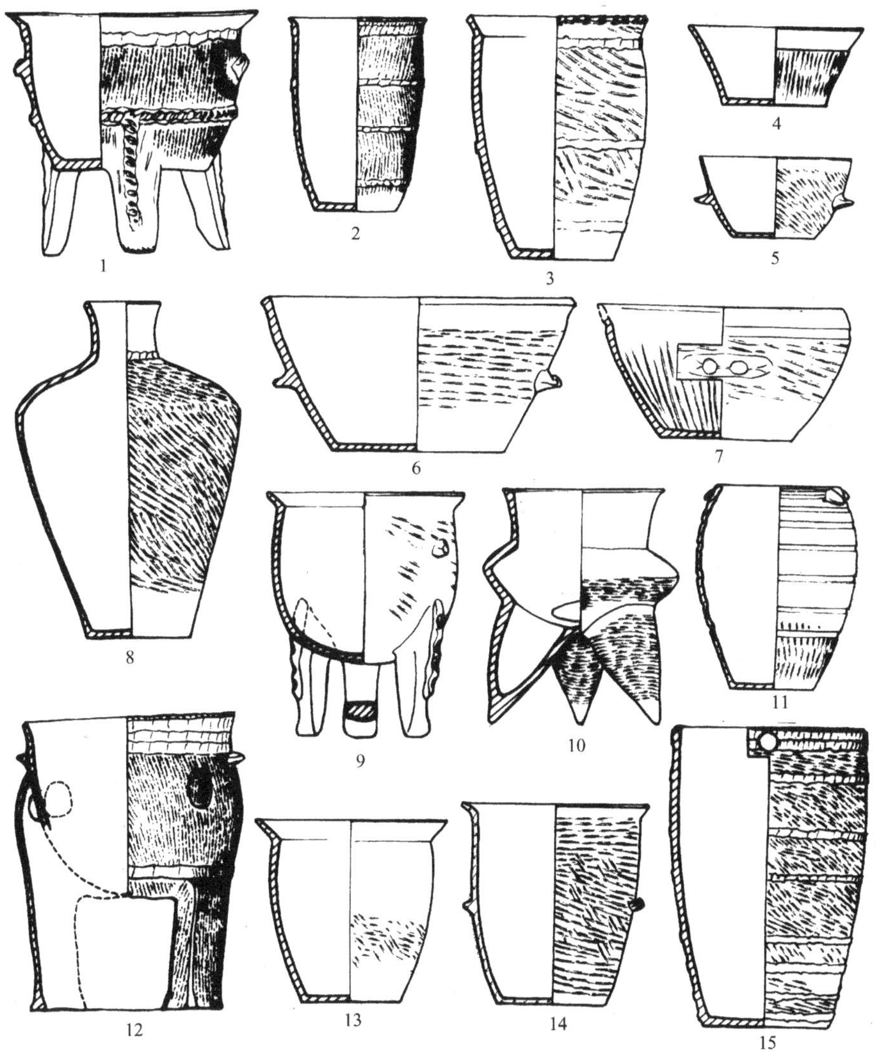

Figure 6-8a. Early Miaodigou II ceramics from the Dongguan site.
1. *penxingding* tripod; 2. *xiaoguan* jar; 3. *daguan* jar; 4. *bo* bowl; 5. *shuangbanbo* bowl; 6. *shuangbanpen* basin; 7. *kecaopen* basin; 8. *xiaokouguan* jar; 9. *guanxingding* tripod; 10. *jia* tripod; 11. *liankouweng* jar; 12. *fuzao* cook stove; 13. dakouguan jar; 14. *shuangbanguan* jar; 15. *gang* jar.

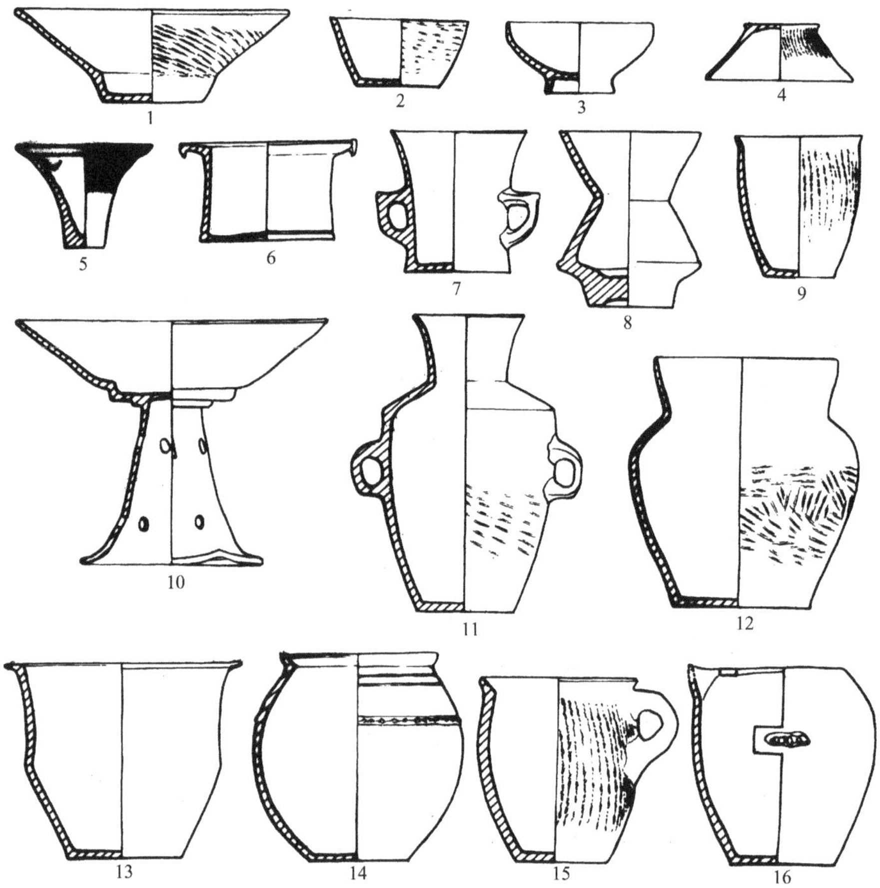

Figure 6-8b. Early Miaodigou II ceramics from the Dongguan site.
1. *aoxinpen* basin; 2. *wan* bowl; 3. *quanzuwan* bowl; 4. *gai* lid; 5. *changkoubei* cup; 6. *tongxingbei* cup; 7. *shuang'erbei* cup; 8. *quanzubei* cup; 9. *weixingguan* jar; 10. *dou* stemmed dish; 11. *ping* bottle; 12. *gaolingguan* jar; 13. *kuanyanpen* basin; 14. *yuanfuguan* jar; 15. *dan'erguan* jar; 16. *dailiuguan* jar.

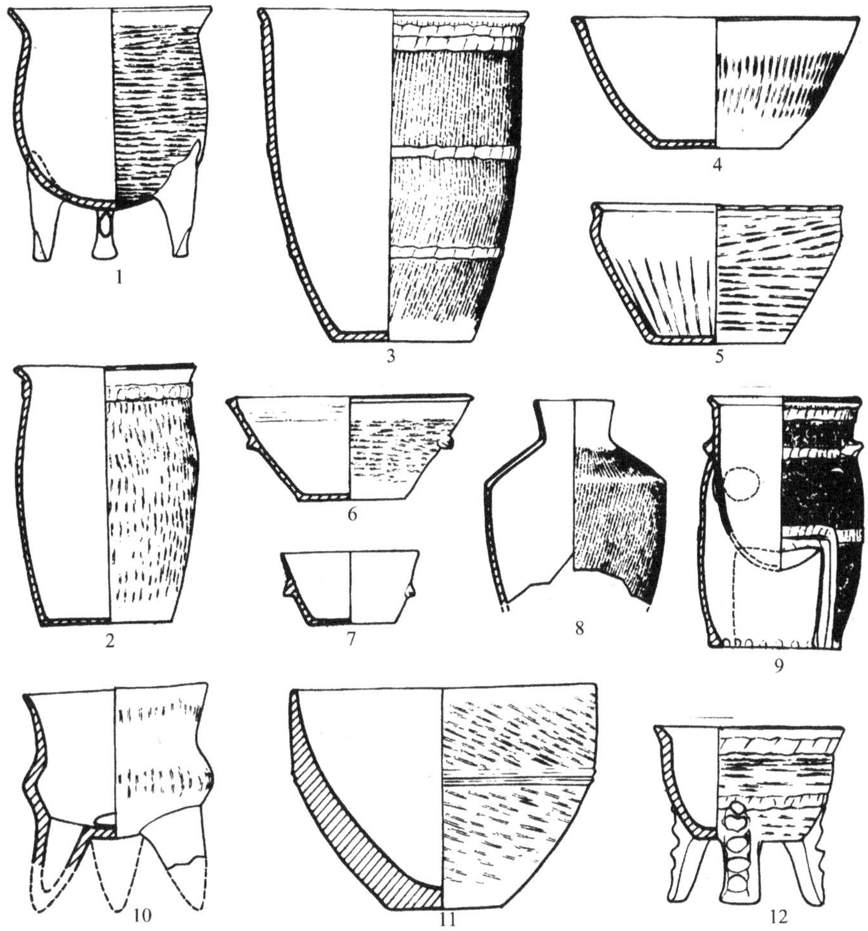

Figure 6-9a. Middle Miaodigou II ceramics from the Dongguan site.
1. *guanxingding* tripod; 2. *xiaoguan* jar; 3. *daguan* jar; 4. *bo* bowl; 5. *kecaopen* basin; 6. *shuangbanpen* basin; 7. *shuangbanbo* bowl; 8. *xiaokouguan* jar; 9. *fuzao* cook stove; 10. *jia* tripod; 11. houtai*gang* jar; 12. *penxingding* tripod.

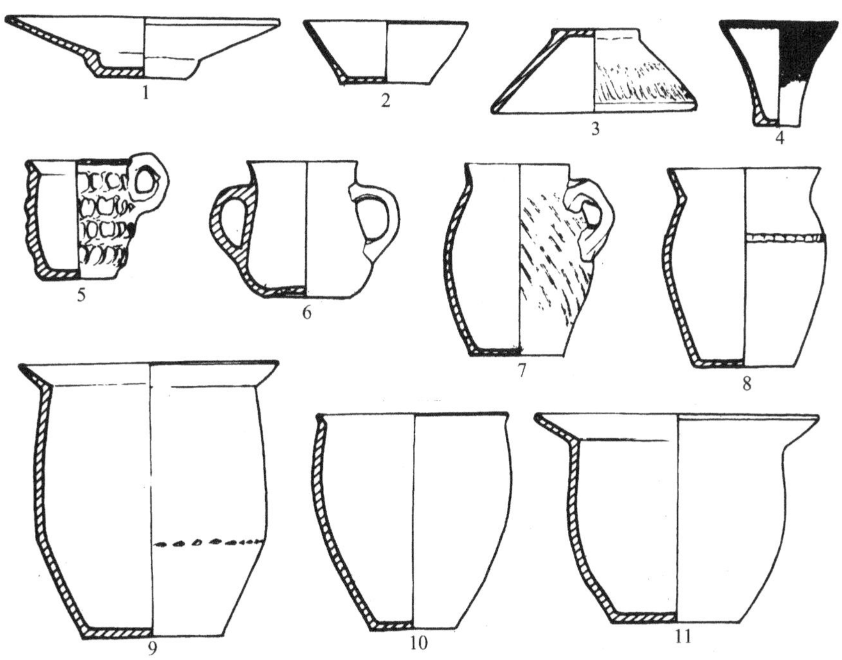

Figure 6-9b. Middle Miaodigou II ceramics from the Dongguan site.
1. *aoxinpen* basin; 2. *wan* bowl; 3. *gai* lid; 4. *changkoubei* cup; 5. *dan'erbei* cup; 6. *shuang'erbei* cup; 7. dan'erguan jar; 8. *kuanyanguan* jar; 9. *dakouguan* jar; 10. *sumianguan* jar; 11. *kuanyanpen* basin.

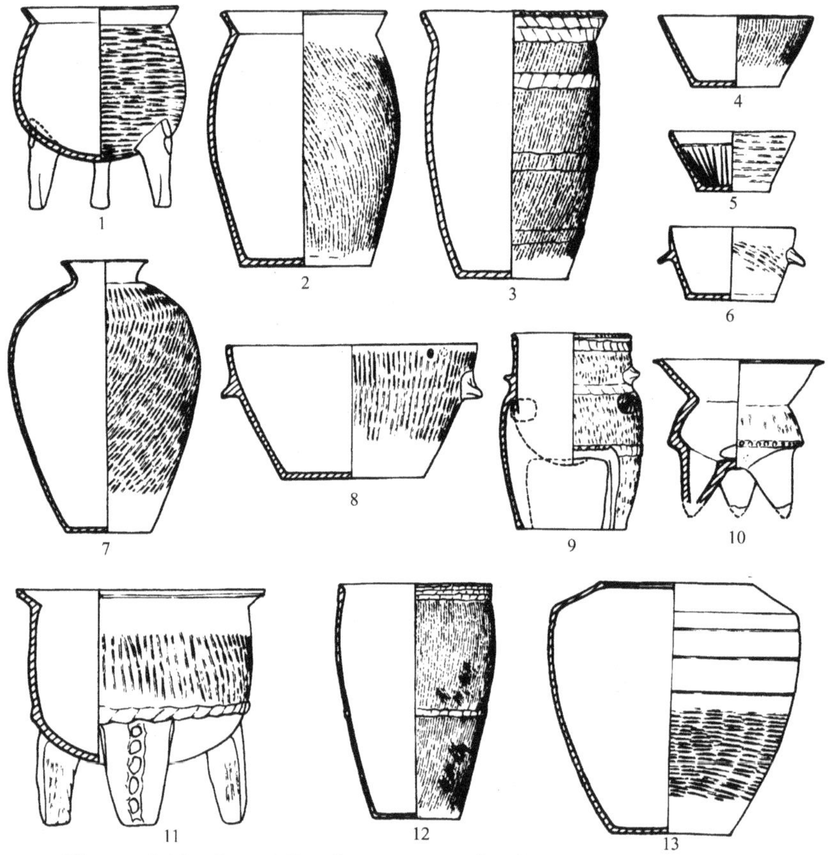

Figure 6-10a. Late Miaodigou II ceramics from the Dongguan site.
1. *guanxingding* tripod; 2. *xiaoguan* jar; 3. *daguan* jar; 4. *bo* bowl; 5. *kecaopen* basin; 6. *shuangbanbo* bowl; 7. *xiaokouguan* jar; 8. *shuangbanpen* basin; 9. *fuzao* cook stove; 10. *jia* tripod; 11. *penxingding* tripod; 12. *gang* jar; 13. *liankouweng* jar.

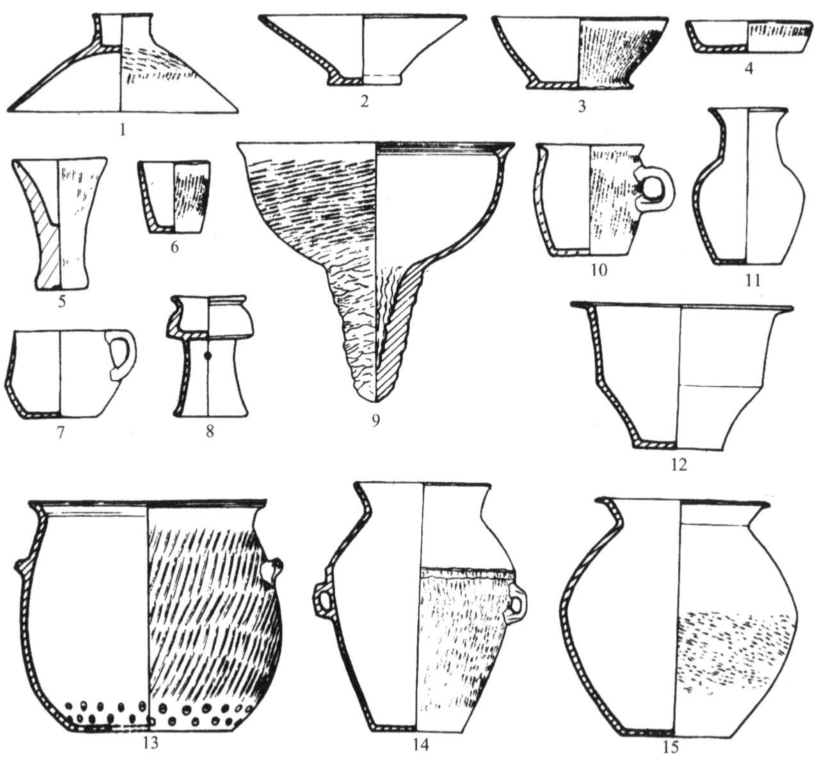

Figure 6-10b. Late Miaodigou II ceramics from the Dongguan site.
1. *gai* lid; 2. *aoxinpen* basin; 3. *houdiwan* bowl; 4. *wan* bowl; 5. *changkoubei* cup; 6. *tongxingbei* cup; 7. *dan'erbei* cup; 8. *quanzubei* cup; 9. *jiandigang* jar; 10. *dan'erguan* jar; 11. *ping* bottle; 12. *kuanyanpen* basin; 13. *zeng* steamer; 14. *shuang'erguan* jar; 15. *yuanfuguan* jar.

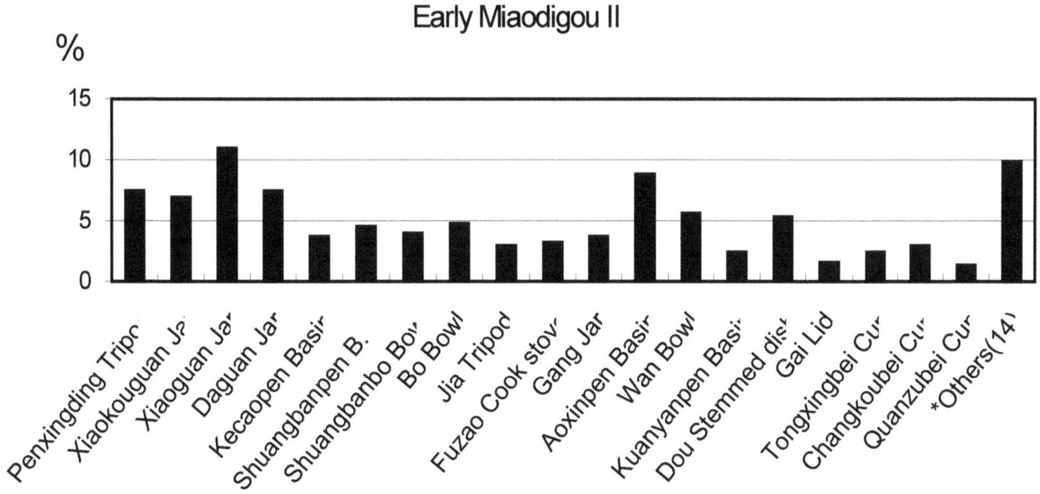

Figure 6-11a. Distribution of ceramic classes in the early Miaodigou II assemblage.
*Others (14): 14 ceramic classes less than five specimens, including the *guanxingding* tripod, *huandigang* jar, *liankouweng* jar, *dan'erguan* jar, *yuanfuguan* jar, *dailiuguan* jar, *dakouguan* jar, *ping* bottle, *quanzuwan* bowl, *shuangbanguan* jar, *gaolingguan* jar, *weixingguan* jar, *shuang'erbei* cup, and *zeng* steamer.

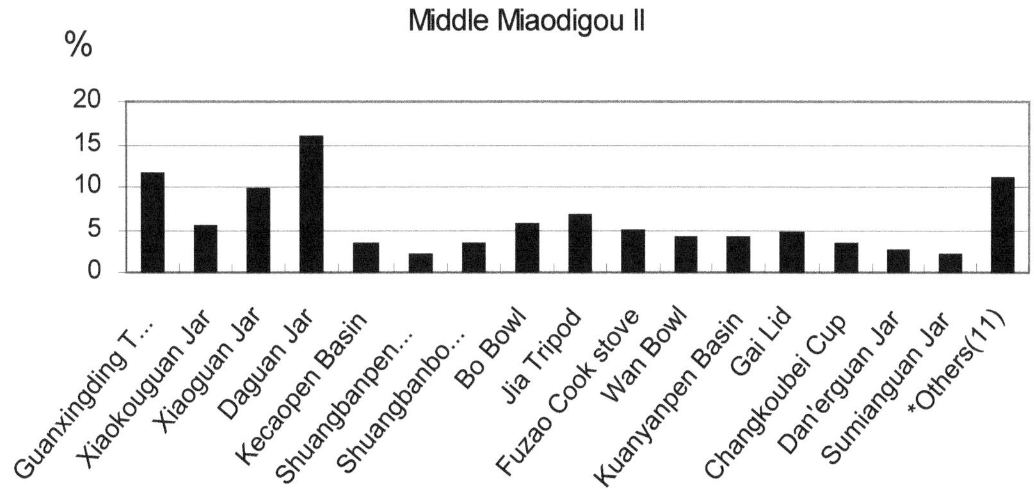

Figure 6-11b. Distribution of ceramic classes in the middle Miaodigou II assemblage.
*Others (11): 11 ceramic classes less than five specimens, including the *penxingding* tripod, *houtaigang* jar, *kuanyanguan* jar, *dakouguan* jar, *aoxinpen* basin, *dou* stemmed dish, *dan'erbei* cup, *shuang'erbei* cup, *tongxingbei* cup, *quanzubei* cup, and *zeng* steamer.

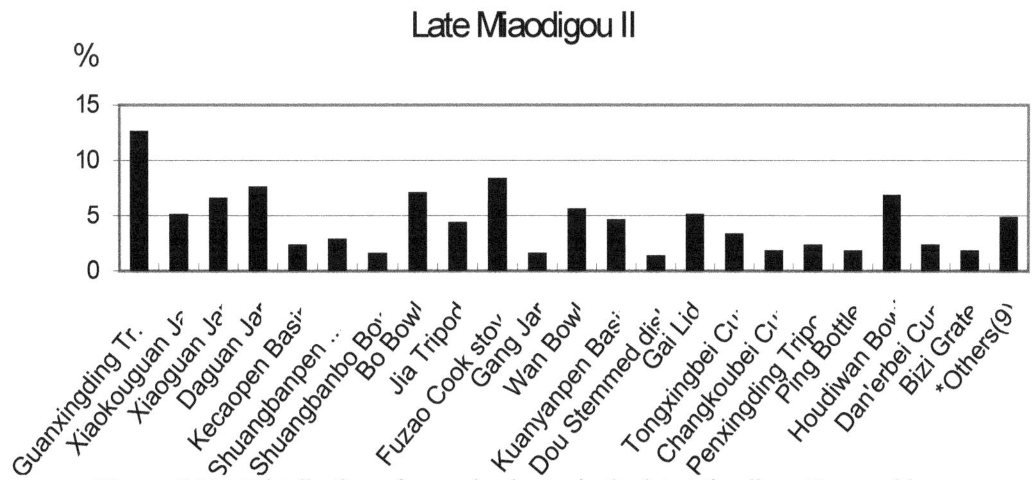

Figure 6-11c. Distribution of ceramic classes in the late Miaodigou II assemblage.
*Others (9): Nine ceramic classes less than five specimens, including the *jiandigang* jar, *liankouweng* jar, *shuang'erguan* jar, *dan'erguan* jar, *yuanfuguan* jar, *dakouguan* jar, *aoxinpen* basin, *quanzubei* cup, and *zeng* steamer.

Table 6-1. Frequency of pottery colours for the three phases of the Miaodigou II period at the Dongguan site.

	Grey and Pale-brown	Red and Reddish-brown	Black
Early	89.5%	10.2%	0.3%
Middle	88.2%	10.2%	1.8%
Late	79.9%	16%	4.1%

Table 6-2. Frequency of primary paste categories for the three phases of the Miaodigou II period at the Dongguan site.

	Sandy paste	Silt paste
Early	67.4%	32.6%
Middle	73.6%	26.4%
Late	74.2%	25.8%

Table 6-3. Frequency of surface treatments and decorations for the three phases of the Miaodigou II period at the Dongguan site.

	Basket-marking	Cord-marking	Check stamping	Clay strip	Bowstring	Plain
Early	25.2%	34%	0.04%	11.63%	0.03%	29.1%
Middle	40.4%	23%	0.3%	6.2%	0.08%	30.02%
Late	44.15%	15.45%	2%	4.6%	0.24%	33.56%

According to the excavation report, for the two primary categories of pottery paste, sandy-paste vessels are far more numerous than silt-paste vessels for the three phases in the Miaodigou II assemblage (Table 6-2) (National Museum of Chinese History et al. 2001:184,247,292). This situation is also opposite to that of the Dongguan I assemblage. By comparison with the geological sand grain chart (see Chapter Four), sandy paste includes medium, coarse, and very coarse sand, and there was no fine-sandy paste in the Miaodigou II assemblage. I measured and recorded the texture of pottery paste for the eight ceramic types mentioned above. As three of them (*kecaopen* basin, *shuangbanpen* basin, *shuangbanbo* bowl) are small in sample size, I focus on the results of the other five types (Appendix 4). For some ceramic classes, such as *ding* tripods, *xiaoguan* jars and *daguan* jars, there is a certain correlation between them and primary paste categories. There is no or few silt paste, and they mainly comprise diverse sandy paste. However, these ceramic classes do not strictly correlate to the secondary paste categories. Generally there are no strong correlations between paste categories and shape classes. Almost each ceramic type with relatively large sample size contains several paste categories (see Appendix 4). This situation is very different from the Dongguan I assemblage (see Chapter Five).

For a single ceramic type in the Miaodigou II assemblage, the major composition of paste texture often varied over time and did not remain stable between the three phases. For example, the paste category with the most samples for the *xiaoguan* jar is respectively coarse, very coarse, and medium sand from the early, middle to late phase; the paste category with the largest sample size for the *bo* bowl is respectively silt, coarse and medium sand from the early, middle to late phase (see Appendix 4). In a word, the texture of pottery paste exhibits high diversity and variability for all three phases.

The surface treatment and decoration in the Miaodigou II assemblage encompasses basket-marking, cord-marking, check stamping, plain, and some clay strips and bowstrings as the secondary decorations. Among them, basket-marking, cord-marking, and plain constitute the main body in this aspect. A general trend is that basket-marking increased, cording-marking and clay strips decreased, and plain maintained relatively stable in number from the early, middle to late phase (Table 6-3) (National Museum of Chinese History et al. 2001:184,247,292).

There are no strong correlations between decorations and ceramic classes for the Miaodigou II assemblage, and most vessel forms have two or more categories of surface treatments and decorations. However, there is usually a major category of surface treatment or decoration for a ceramic type. For most ceramic types, basket-marking is evidently the main form of body decorations. In some circumstances, cord-marking constitutes the main body of decorations, such as for *xiaoguan* jars in middle and late phase, and *daguan* jars in the late phase. In some other circumstances, most vessels are plain, such as *bo* bowl in the late phase. For most ceramic classes, the major composition of surface treatments and decorations sustained coherent from the early, middle to late phase, but there were still variations for some pottery types, such as *xiaoguan* jar, *daguan* jar, and *bo* bowl (Appendix 5). In brief, the surface treatment and decoration during the Miaodigou II period displayed relatively high diversity and variability, but the degree of standardisation in this aspect seems to be higher than that in pottery paste.

The manufacturing techniques for Miaodigou II pottery were much more complicated than those for Dongguan I pottery. A vast majority of vessels were still built by hand (coiling), but meanwhile moulding and throwing techniques appeared. Moulding was mainly used for legs of *jia* tripods, which were formed by combination of coiling and moulding. Wheel-throwing was not yet popular, and it was only found on a few types of small vessels, such as *changkoubei* cups, *quanzubei* cups and some bowls (Li 2001). Shaping and finishing techniques still mainly included beating, scraping and burnishing. Besides, the "slow-wheel" technique was increasingly used for shaping and smoothing of several ceramic types, including the *ding* tripod, *xiaokouguan* jar, *daguan* jar, and s*huangbanpen* basin (see Fig. 6-8~ 6-10).

Overall, there were certain but not strong correlations between manufacturing techniques and ceramic classes, and there were no significant changes in manufacturing techniques between the three phases. The emergence of some special techniques, such as moulding, was mainly associated with the form and function of the vessels, and indicated the advance of manufacturing techniques and improvement of standardisation of these vessels.

In summary, the nominal variables in the Miaodigou II assemblage display high diversity and relatively low standardisation for all three phases. This possibly related to many independent producers and less restriction to manufacture, and reflected lower specialisation. Competitive emulation among individual workshops or craftsmen, and various social demands may have been the major reasons leading to pottery diversity.

6.4.2. Analyses of Metric Data

As in Chapter Five, I will use the coefficient of variation (CV) and analysis of variance (ANOVA, F ratio) for metric data to investigate standardisation and variability of ceramic vessels for the three phases during the Miaodigou II period, and then assess the degree of specialisation of pottery production.

As mentioned at the beginning of this section (**6.4.**), I selected eight pottery types to measure and record metric variables. However, the sample sizes of four of them in each phase are too small to be suitable for statistical analyses, so I have to focus on the other four ceramic classes: *ding* tripods, *xiaoguan* jars, *daguan* jars, and *bo* bowls. It is notable that there are two types of *ding* tripods in the Miaodigou II assemblage: *penxingding* tripods and *guanxingding* tripods (see Fig. 6-8~6-10). The former made up the main body of *ding* tripods in the early phase, and the latter composed the main body of *ding* tripods in the middle and late phase. The statistical data for *ding* tripods refer to *penxingding* tripods in the early phase, and to *guanxingding* tripods in the middle and late phase.

For the foregoing four classes, most ceramic types in a specific chronological phase are larger than 20 in sample size, but many data sets for single variables are less than 20 in number, below the minimum sample size for the statistical analyses under the optimal condition in this study (see Chapter Four). This is because a large number of sherds other than whole vessels exist as pottery specimens for each ceramic category. I keep the CV scores where there are more than five specimens, and reject those CV scores with less. I will take advantage of CV values where there are more than 20 specimens (especially CV values for pooled data) to examine the significance of CV values of smaller sets. In addition, fortunately, most pooled data from the three phases for the eight ceramic classes are more than or around 20 in number, so I give CV values for all pooled data of the eight ceramic classes mentioned above, which will be used for comparison between the assemblages.

As mentioned in Chapter Four, I measured and recorded nine metric variables for all simple and complex vessels in shape in this study. For the width of five marks, I only analyse statistical data of basket-marking because it constitutes the main body of decorations for most ceramic categories in the Miaodigou II assemblage. Among other size variables, only *ding* tripods and jar vessels address rim length, neck diameter and body width, and only *ding* tripods address body height and leg length. These five dimensions are not relevant to bowls and basins. Moreover, there are always strong correlations for orifice diameter, neck diameter, and body width, and there are also always strong correlations between vessel height and body height (Table 6-4). It is not necessary to use all these dimensions for statistical analyses. Here I only show the statistical data of six size variables: orifice diameter, base diameter, vessel height, and width of five marks (for all ceramic classes); body width and rim length (for tripods, jars). In addition, the ratio orifice diameter/vessel height is still an important variable for the analyses of standardisation or variability.

Table 6-4. Correlation coefficients among metric variables for the three ceramic classes in the Miaodigou II assemblage of the Dongguan site.

		*OD/ND	*OD/BW	*VH/BH
Ding Tripod	Early	0.990	0.966	0.949
	Middle	0.981	0.898	0.990
	Late	0.956	0.839	0.974
Xiaoguan Jar	Early	0.971	0.850	
	Middle	0.928	0.724	
	Late	0.989	0.745	
Daguan Jar	Early	0.959	0.967	
	Middle	0.955	0.928	
	Late	0.942	0.925	

*OD/ND: Orifice diameter/Neck diameter. OD/BW: Orifice diameter/Body width.
VH/BH: Vessel height/Body height.

Firstly, I will consider the variation between the three phases through the analysis of variance (ANOVA) for the four major ceramic classes (Table 6-5). Due to the small samples (less than six specimens for a data set), a few variables of several vessel forms have to be rejected from this analysis, including width/5 marks for the *xiaoguan* jar and the *bo* bowl, body width, vessel height and the ratio orifice diameter/vessel height for the *daguan* jar. For the *ding* tripod, ANOVA in Table 6-5 only addresses *guanxingding* tripods of the middle and late phase, which are very different from *penxingding* tripods of the early phase. It is noteworthy that there is no one ceramic type in which all size variables vary significantly over time at the 0.05 level for ANOVA. Among six variables for *ding* tripods, only two of them vary significantly ($p<0.05$). There are significant variations for five variables of *xiaoguan* jars among the six available variants, three variables of *daguan* jars among the four available variants, and two variables of *bo* bowls among the four available variants (see Table 6-5). Apparently, there is certain statistically significant difference in variability between the three phases for each ceramic type in some but not all variables. These variations are embodied more obviously in *xiaoguan* and *daguan* jars. However, apart from *bo* bowls, there is usually no significant difference for the ratio of orifice diameter-to-vessel height of other shape classes after the size problem is eliminated by the ratio ($p>0.05$). In conclusion, some significant variations for pottery dimensions had occurred between the three phases during the Miaodigou II period, but they are limited.

In the following analyses, I will firstly assess the variations of CV values for the four major ceramic classes between the three phases, and then evaluate the significance of CV values for the pooled data of the eight ceramic classes during the whole Miaodigou II period (Table 6-6). A marked phenomenon is that there are obvious differences for CV values among variants for these ceramic classes. Comparatively, CV values for rim length are the highest, the second is those for width of five marks, and CV values for the ratio orifice diameter/vessel height are the lowest, whereas CV values for orifice diameter, body width, base diameter, and vessel height relatively approximate one another. I will treat CV values of these distinct variables respectively in the following analyses.

There is no single common tendency for CV scores of the four major ceramic classes from the early to late phase. For *ding* tripods, CV values for orifice diameter, body width, height, and rim length tend to be increasingly low, but there is no such kind of trend for CV values of width of five marks and the ratio orifice diameter/vessel height. For *bo* bowls, CV values for orifice diameter and base diameter also become increasingly low, but for other variables this is not the case. For *xiaoguan* jars, CV values for rim length and the ratio orifice diameter/vessel height tend to be increasingly high, but there are no regular changes for other variants. For *daguan* jars, only CV values for base diameter become increasingly high, but for other variables this is not the case (see Table 6-6). In brief, there are clearly some variations for every shape class in diverse variables between the three phases. Nevertheless, the variations for these pottery categories are limited. There is therefore no common and identifiable trend of standardisation/variability from the early to late phase among these pottery types. These conclusions are consistent with the analytical results of ANOVA and nominal data.

Table 6-5. Analysis of variance (ANOVA) for the four classes of ceramics between the three phases in the Miaodigou II assemblage of the Dongguan site.

		Orifice diameter	Body width	Base diameter	Vessel height	Rim length	*Width/ 5M(BM)	*O/H
Ding Tripod	F ratio	9.024	0.166		<0.001	7.722	0.330	0.358
*(M&L)	P	0.004	0.685		0.987	0.007	0.568	0.553
Xiaoguan Jar	F ratio	3.240	7.061	12.747	4.329	4.368	__	1.198
	P	0.045	0.002	<0.0001	0.020	0.016	__	0.313
Daguan Jar	F ratio	8.417	__	5.398	__	2.710	8.326	__
	P	0.001	__	0.004	__	0.073	0.001	__
Bo Bowl	F ratio	1.904		0.329	6.874		__	4.192
	P	0.159		0.721	0.002		__	0.020

Note: Significant at $P<0.05$.
*Width/5M(BM): Width/5 marks (Basket-marking). O/H: Orifice diameter/Vessel height.
(M&L): (Middle & Late phase).

Table 6-6. Sample means (M), standard deviations (SD), and coefficients of variation (C.V.) for the eight classes of ceramics in the Miaodigou II assemblage of the Dongguan site.

		Orifice diameter	Body width	Base diameter	Vessel height	Rim length	*Width/ 5M(BM)	*O/H
Ding Tripod								
Early	n=28	20	15	14	10	19	8	10
	M(cm)	21.62	19.14	13.38	22.93	2.82	2.28	1.06
	SD	6.30	4.72	3.54	4.67	1.02	0.60	0.18
	C.V.	**29.13%**	**24.66%**	**26.45%**	**20.37%**	**36.14%**	**26.49%**	**17.26%**
Middle	n=28	21	12		8	21	13	8
	M(cm)	19.31	17.67		21.99	3.17	3.37	0.75
	SD	5.22	2.55		3.06	1.08	0.66	0.07
	C.V.	**27.06%**	**14.44%**		**13.93%**	**34.16%**	**19.52%**	**9.21%**
Late	n=50	49	36		11	49	38	11
	M(cm)	16.31	17.34		22.01	2.61	3.08	0.76
	SD	3.07	2.34		2.74	0.58	0.74	0.07
	C.V.	**18.84%**	**13.49%**		**12.47%**	**22.26%**	**24.20%**	**9.72%**
Combined	n=78	70	48		19	70	51	19
*(M&L)	M(cm)	17.21	17.42		22.00	2.78	3.15	0.75
	SD	4.05	2.37		2.80	0.80	0.73	0.07
	C.V.	**23.54%**	**13.61%**		**12.72%**	**28.78%**	**23.12%**	**9.28%**
Xiaoguan Jar								
Early	n=41	37	32	31	27	39	21	27
	M(cm)	15.39	15.72	9.58	21.00	1.91	2.51	0.72
	SD	3.12	3.03	1.46	4.56	0.45	0.72	0.09
	C.V.	**20.24%**	**19.26%**	**15.27%**	**21.70%**	**23.72%**	**28.61%**	**12.89%**
Middle	n=24	18	11	14	7	18	5	7
	M(cm)	16.36	18.67	12.10	25.47	2.04		0.69
	SD	2.22	3.33	2.09	3.50	0.54		0.09
	C.V.	**13.57%**	**17.81%**	**17.27%**	**13.76%**	**26.35%**		**13.19%**
Late	n=26	23	14	10	7	23	7	7
	M(cm)	17.28	19.06	11.64	25.24	2.37	3.10	0.66
	SD	2.70	3.37	1.86	5.06	0.79	0.83	0.11
	C.V.	**15.62%**	**17.69%**	**15.97%**	**20.06%**	**33.34%**	**26.91%**	**15.91%**
Combined	n=91	78	57	55	41	80	33	41
	M(cm)	16.17	17.11	10.60	22.49	2.07	2.71	0.71
	SD	2.89	3.50	2.05	4.85	0.61	0.79	0.10
	C.V.	**17.89%**	**20.43%**	**19.33%**	**21.59%**	**29.46%**	**29.29%**	**13.49%**
Daguan Jar								
Early	n=28	26	16	13	12	26	11	11
	M(cm)	27.48	26.95	14.40	35.10	2.65	2.49	0.74
	SD	3.54	3.06	1.37	2.89	0.63	0.28	0.05
	C.V.	**12.88%**	**11.35%**	**9.52%**	**8.23%**	**23.90%**	**11.18%**	**6.82%**
Middle	n=39	29	10	16	7	30	6	6
	M(cm)	31.81	32.02	17.53	39.53	2.74	3.00	0.79

		Orifice diameter	Body width	Base diameter	Vessel height	Rim length	*Width/ 5M(BM)	*O/H
	SD	4.24	5.49	2.74	4.68	0.84	0.30	0.08
	C.V.	**13.33%**	**17.16%**	**15.64%**	**11.83%**	**30.80%**	**10.02%**	**10.32%**
Late	n=30	25	5	8	2	24	17	2
	M(cm)	29.59		17.95		3.16	3.40	
	SD	3.87		3.02		0.96	1.04	
	C.V.	**13.07%**		**16.80%**		**30.27%**	**30.55%**	
Combined	n=97	80	31	37	21	80	34	19
	M(cm)	29.71	29.66	16.52	36.80	2.84	3.04	0.76
	SD	4.26	4.85	2.84	4.14	0.84	0.85	0.06
	C.V.	**14.33%**	**16.36%**	**17.18%**	**11.25%**	**29.51%**	**28.13%**	**8.54%**
Bo Bowl								
Early	n=18	16		16	16		9	14
	M(cm)	16.57		9.13	9.06		2.16	1.97
	SD	4.32		2.97	2.55		0.55	0.36
	C.V.	**26.07%**		**32.54%**	**28.18%**		**25.60%**	**18.07%**
Middle	n=14	14		14	14		4	14
	M(cm)	16.01		9.09	7.03			2.34
	SD	3.64		2.30	2.06			0.44
	C.V.	**22.76%**		**25.26%**	**29.27%**			**18.78%**
Late	n=28	28		28	28		7	28
	M(cm)	14.60		8.61	6.68		2.79	2.24
	SD	2.65		1.86	1.83		0.30	0.31
	C.V.	**18.16%**		**21.60%**	**27.33%**		**10.72%**	**13.98%**
Combined	n=60	58		58	58		20	56
	M(cm)	15.49		8.87	7.42		2.50	2.20
	SD	3.47		2.29	2.31		0.56	0.38
	C.V.	**22.41%**		**25.77%**	**31.10%**		**22.53%**	**17.22%**
Kecaopen Basin								
Combined	n=31	30		25	24		15	24
	M(cm)	18.60		11.56	9.96		3.07	1.83
	SD	5.65		2.82	3.56		0.81	0.29
	C.V.	**30.39%**		**24.41%**	**35.76%**		**26.25%**	**15.57%**
Shuangbanpen Basin								
Combined	n=33	33		21	21		26	21
	M(cm)	29.05		15.27	15.14		3.04	2.04
	SD	4.53		1.64	2.26		0.91	0.44
	C.V.	**15.60%**		**10.72%**	**14.90%**		**29.97%**	**21.51%**
Shuangbanbo Bowl								
Combined	n=29	29		22	22		12	22
	M(cm)	16.76		9.20	8.74		2.32	1.92
	SD	2.49		2.33	1.98		0.65	0.33
	C.V.	**14.86%**		**25.27%**	**22.69%**		**27.84%**	**16.95%**

		Orifice diameter	Body width	Base diameter	Vessel height	Rim length	*Width/ 5M(BM)	*O/H
Xiaokouguan Jar								
Combined	n=59	48	17	15	11	46	26	11
	M(cm)	11.41	28.88	14.15	42.97	6.50	2.63	0.26
	SD	1.55	5.65	1.39	4.90	1.50	0.77	0.04
	C.V.	**13.54%**	**19.55%**	**9.81%**	**11.40%**	**23.02%**	**29.28%**	**14.29%**

*Width/5M(BM): Width/5 marks (Basket-marking). O/H: Orifice diameter/Vessel height.
(M&L): (Middle & Late phase).

For these four ceramic categories in all three phases, the whole variable range of CV values for orifice diameter, body width, base diameter and vessel height is from 8.23% to 32.54%, and most of them vary between 10% and 28%. The whole variable range of CV values for the ratio orifice diameter/vessel height is from 6.82% to 18.78%, and most of them vary between 9% and 17%. The whole variable range of CV values for rim length is from 22.26% to 36.14%, and most of them range from 22% to 34%. The whole variable range of CV values for width of five marks is from 10.02% to 30.55%, and most of them range from 19% to 30% (see Table 6-6). The variable ranges of CV values for orifice diameter, base diameter, vessel height, and the ratio orifice diameter/vessel height between the three phases in the Miaodigou II period are similar to those corresponding variable ranges in the Dongguan I period (see Chapter Five). Likewise, the ratio of orifice diameter-to-vessel height also shows commonly lower CV values than all single dimensions in the Miaodigou II assemblage, and is the more conservative and effective indicator of standardisation/variability. Besides, CV values for rim length are commonly higher than CV values of other variables. This pattern indicates that the rim length is the most flexible among all vessel dimensions.

Turning to CV values for pooled data of eight vessel forms in the Miaodigou II assemblage. The whole variable range of CV values for orifice diameter, body width, base diameter and vessel height is from 9.81% to 35.76%, and most of them range from 10% to 25%. The whole variable range of CV values for the ratio orifice diameter/vessel height is from 8.54% to 21.51%, and most of them vary between 13% and 17%. These patterns are also similar to those in the Dongguan I assemblage, and the variation spectrum of the former is even a little bigger than the latter (see Chapter Five). Likewise, the rim length has commonly high CV values, ranging from 23.02% to 29.51%. CV values for width of five marks are similar in this case, ranging from 22.53% to 29.98% (see Table 6-6). In short, CV values for combined data in the Miaodigou II assemblage are still relatively high, indicating relatively low standardisation and specialisation.

The sample sizes for most variants of pooled data in these eight vessel forms are more than 20, so CV values for pooled data may be more reliable. Similar to Dongguan I, most CV values for combined data of the four major ceramic types (the *ding* tripod, *xiaoguan* jar, *daguan* jar, and *bo* bowl) in the Miaodigou II assemblage fall between the lowest and highest values of the three phases. Only a few CV values for pooled data are a little higher than those in the three phases, such as CV values for body width and base diameter of *xiaoguan* jars, orifice diameter and base diameter of *daguan* jars, and vessel height of *bo* bowls (see Table 6-6). These patterns suggest that most CV values of those variables with small samples (less than 20) do not seem to excessively deviate from expected values. In the Miaodigou II assemblage, the variability of CV values seems to be related to ceramic classes, metric variables, and chronological phases much more rather than sample size.

In conclusion, we can draw two important results from the analyses of metric data. Firstly, although there are some significant variations in some variables for many shape classes between the three phases, these are limited. There is not a common and clear trend of standardisation/variability from the early to late phase among these pottery types. The changes between the three phases are probably due to technological or other elements, rather than the variance of productive modes. Secondly, the variable ranges of CV values among the three phases and pooled data in the Maiodigou II assemblage are similar to those in the Dongguan I assemblage, and most CV values during this period are still relatively high, indicating lower uniformity and standardisation. These conclusions are similar to the analytical results of nominal data. These observations suggest that the degree of specialisation in the Miaodigou II period may have been similar to that in the Dongguan I period. Pottery production during the Miaodigou II period was also probably the small-scale and part-time activity.

6.5. Summary

On the basis of the foregoing analyses of direct and indirect evidence, we can outline and summarise some essential characteristics of pottery production in a chiefdom-level society in the Yuanqu Basin (focusing on the Miaodigou II settlement at the Dongguan site).

There is much direct evidence of pottery production in many sites of the Yuanqu Basin for this period, so at the regional level the pottery production during the Miaodigou II period can be seen to have been dispersed rather than concentrated. The kilns and relevant facilities (including working areas, storage pits, structures, and pits for clay preparation) at Dongguan and Ningjiapo possibly cover a full set of procedures for pottery making, firing, and storage, so specialised workshops for pottery production may have appeared in the Yuanqu Basin. It is very likely that individual workshop industry was the major mode of production in this period, though household industry could not yet be ruled out completely. Accordingly, there may have been some similarities in the mode of production between the Miaodigou II and Dongguan I period. This conclusion can be further supported by CV values for metric variables between them. The variable ranges of CV values for the common variants between them are similar to each other. Even if some kind of change in productive modes had happened between the two periods, such as from household industry to individual workshops, this change can hardly be visible in CV values, since the difference between these two modes of production is very fine and they are similar in many aspects, as discussed in previous two chapters. Therefore, it seems that the development of specialisation was limited at the time when the social form had developed from the simple to complex society in the Yuanqu Basin.

On the other hand, some variations did occur during this period. In addition to the appearance of workshops exhibited in direct evidence, the nominal data (including pottery classification) in the Miaodigou II assemblage display higher diversity than those in the Dongguan I assemblage. This probably reflected many independent producers and less restriction on pottery manufacture. Meantime, this situation might imply increasingly intense competition among individual workshops or producers. The lack of prestige pottery items and associations between elite structures or house groups (kin groups) and productive facilities can also support a social context of independent production.

The pottery production indicated by direct evidence was probably small-scale in individual workshops. Nominal and metric data exhibit high diversity/variability, and low standardisation and specialisation, which also support the small-scale and part-time production during the Miaodigou II period at the Dongguan site.

All production facilities of the Dongguan site discussed in this chapter belong to the early phase, so the modes of production in the middle and late phase could not be identified through direct evidence. However, indirect evidence (including nominal and metric data) definitely suggests that variations between the three phases were insignificant. There was not a common identifiable trend of standardisation/variability from the early to late phase among those pottery variables. The variations between the three phases probably resulted from technological or other elements, rather than the changes in the productive mode. The mode of pottery production may have remained unchanged during the Miaodigou II period at the Dongguan site.

Chapter Seven

Pottery Production in a State-level Society: A Case Study of the Erlitou &Erligang Period at the Nanguan Site

7.1. Introduction

As discussed in Chapter Three, during the Erlitou and Erligang period (the Early Bronze Age, ca. 2000/1900-1300 B.C.), the regional polities in the Yuanqu Basin had been integrated into two successive state systems, whose heartlands were located in central Henan at the south of the Yuanqu Basin. This chapter will discuss the mode of pottery production in the Late Erlitou (ca. 1800/1700-1600/1500 B.C.) and Erligang (ca. 1600-1300 B.C.) periods in the Basin, mainly based on the data from the Nanguan site.

No Early Erlitou remains were discovered at Nanguan. During the Late Erlitou period, although Nanguan was probably not the central place in the Basin, it was the second largest and very important settlement with fortification (surrounded by defensive moats). During the Erligang period, Nanguan became the central place in the Basin, and was encircled by a rammed-earth enclosure. The excavation report cannot further partition the Late Erlitou remains of Nanguan chronologically, so all Late Erlitou ceramics will be regarded as one assemblage (spanning around 200 years). The Erligang remains at Nanguan can be subdivided into two chronological phases: the Lower and Upper Erligang (each spanning about 150 years), so the pottery assemblage of Erligang includes two subsets (National Museum of Chinese History et al. 1996).

I will investigate the variance of pottery production between Late Erlitou, Lower Erligang and Upper Erligang, and reconstruct their modes of production through direct and indirect evidence from Nanguan.

7.2. Direct Evidence of Pottery Production

Of five Erlitou sites excavated in the Yuanqu Basin, only two have direct evidence of pottery production. Kiln(s) have been found at Guanjia (Henan), but no details of their number and distribution have been published (Fan 2000). Some pottery anvils or beaters have been discovered at Nanguan (Fig. 7-1) (National Museum of Chinese History et al. 1996, 1997a), but no production facilities, such as kilns, were uncovered at this site. No direct evidence of pottery production has been found at other Erlitou sites, including Fencun, Xiaozhao and Ningjiapo.

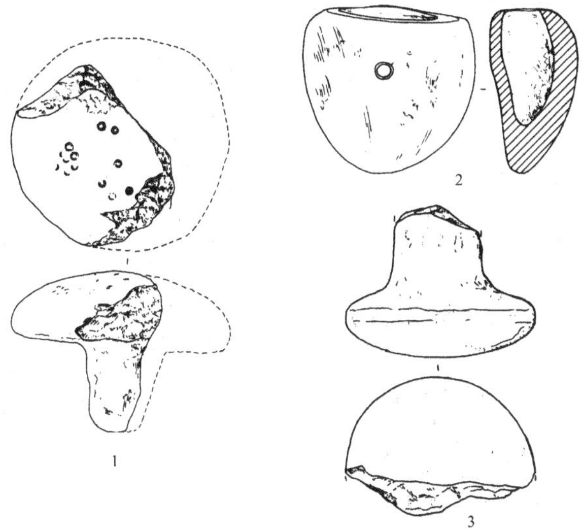

Figure 7-1. Pottery-manufacturing tools of the Late Erlitou and Erligang periods from the Nanguan site.
1. beater (Late Erlitou); 2. anvil (Late Erlitou);
3. beater (Upper Erligang).

Only two Erligang sites have been excavated in the Basin. The Erligang remnants at Ningjiapo have not been reported. I visited this site several times during the excavations: no direct evidence of pottery production, such as manufacturing tools and kilns, were discovered here. Direct evidence of pottery production is abundant at Nanguan, but mostly concentrated in the upper Erligang phase, including 11 kilns and a few anvils and beaters (Fig. 7-1). Only one kiln was recovered in the Lower Erligang phase (National Museum of Chinese History et al. 1996, 1997b).

No other direct evidence of pottery production has been found at any Erlitou and Erligang sites surveyed in the Basin.

This suggests that the pottery production during the Erlitou and Erligang periods was probably concentrated in one, two or a few settlements in the Basin, rather than dispersed over the whole landscape as in the two former periods.

In the light of the sporadic clues of direct evidence mentioned above, we know that during the Late Erlitou and Lower Erligang periods pottery was produced at Nanguan. Due to the lack of evidence of production facilities, it is impossible to reconstruct the modes of production through direct evidence. Therefore, I will focus on the direct evidence of the Upper Erligang period, and then infer the modes of production of Late Erlitou and Lower Erligang at Nanguan by combining other evidence.

Eleven Upper Erligang kilns have been uncovered at Nanguan. Only one kiln (Y6) has been reported (National Museum of Chinese History et al. 1997b). Data of other kilns were obtained from the excavation records of the Gucheng archaeological team. These kilns range from one to 1.5 meters in base diameter of kiln rooms. They are still small in size, but a little bigger than those of Miaodigou II at Dongguan (see Chapter Six). They have many similar properties in structure and style as the kilns of Miaodigou II at Dongguan and Ningjiapo, and many of them were linked to large pits and built into the pit walls.

The kiln Y6 consists of the fire course, fire box, and firing chamber (National Museum of Chinese History et al. 1997b:11,13). The upper section of the firing chamber has been destroyed, so the shape of the top opening of the firing chamber is unknown. The firing chamber is slightly bell-shaped, measuring 1.15 m in base diameter, and 0.6 m in remaining height. The fire box is linked to a pit functioning as the operating area through the fire course. The operating pit is an irregular oval in shape, measuring 1.38 m in maximum diameter and 1.2 m in depth (Fig. 7-2). There is a large amount of firing residue in the pit, such as ash, charcoal, burnt clay, and fragmentary sherds.

Y9 is another well-preserved and elaborate kiln (Fig. 7-3). In addition to the fire course, fire box, and firing chamber, this kiln has a side door connecting with the firing chamber above the fire course. As with Y6, the top of the kiln may have been destroyed. The firing chamber is also bell-shaped, measuring 1.5 m in base diameter, and 0.9 m in remaining height. The fire box is divided into two equal sections by a wall. A notable trait of this kiln is that the entire fire box is formed directly beneath the firing chamber, so that the fuel could be utilised more effectively and the heat could be absorbed more sufficiently by the firing chamber. This kind of structure no doubt indicates progress in firing technology. The kiln Y9 was built immediately beside a large pit, and its side door and fire course connect the pit. The pit is irregular in shape, measuring more than 5 m in length, 4 m in width, and 1.9 m in depth. The kiln is situated at the north-west of the pit. At the south-east portion of the pit is a long slope that likely functioned as the step connecting the outside ground surface. There is a ball-shaped hollow dug into the southern wall of the pit. This measures 1.35 m in maximum diameter and 1.5 m in height. A trapezoidal chamber was built into the western wall of the pit, measuring about 1 m^2 in area. The hollow and chamber could be used as the temporary storage facilities for pottery, tools or other items. It is also likely that they were formed by procuring clay from the pit wall, as showed by the pit IIIH22 of Miaodigou II at Dongguan (see Chapter Six). This pit contained two layers of deposit. The lower deposit is mainly ash, measuring around 0.5 m in thickness, which was likely to be formed during the firing. The upper deposit is mainly reddish-brown earth, measuring around 1.4 m in thickness, which may have been formed after abandonment of the kiln. At the centre of the pit and under the upper deposit is there a grave dug into the lower deposit. It is possible that this

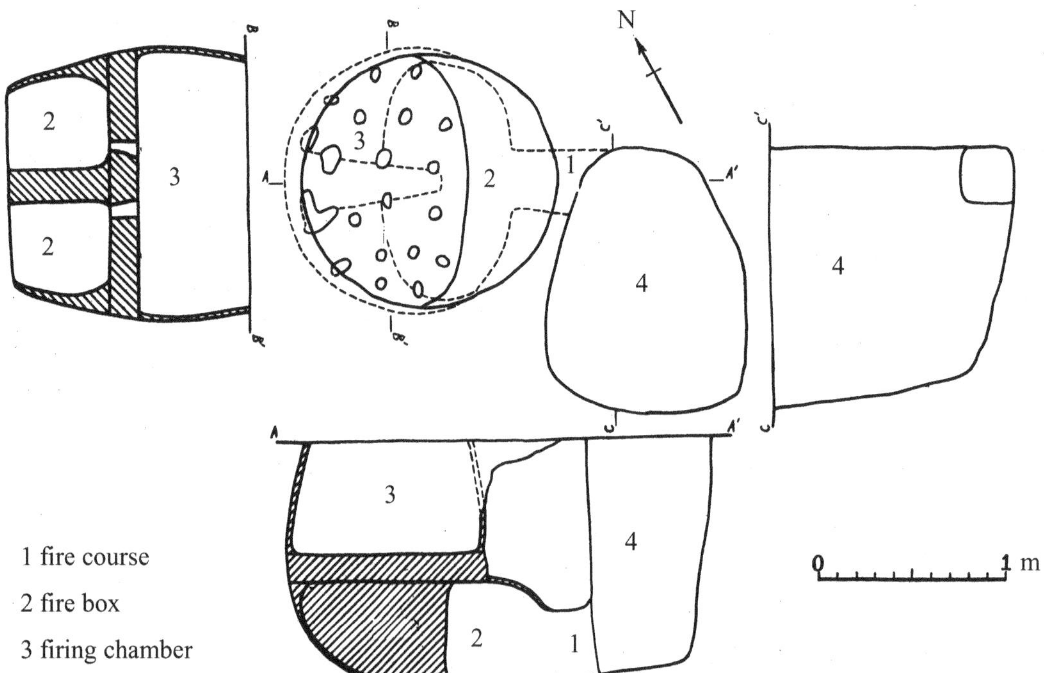

Figure 7-2. Plan and cross-sections of the kiln Y6 at Nanguan (after National Museum of Chinese History et al. 1997b:fig.16)

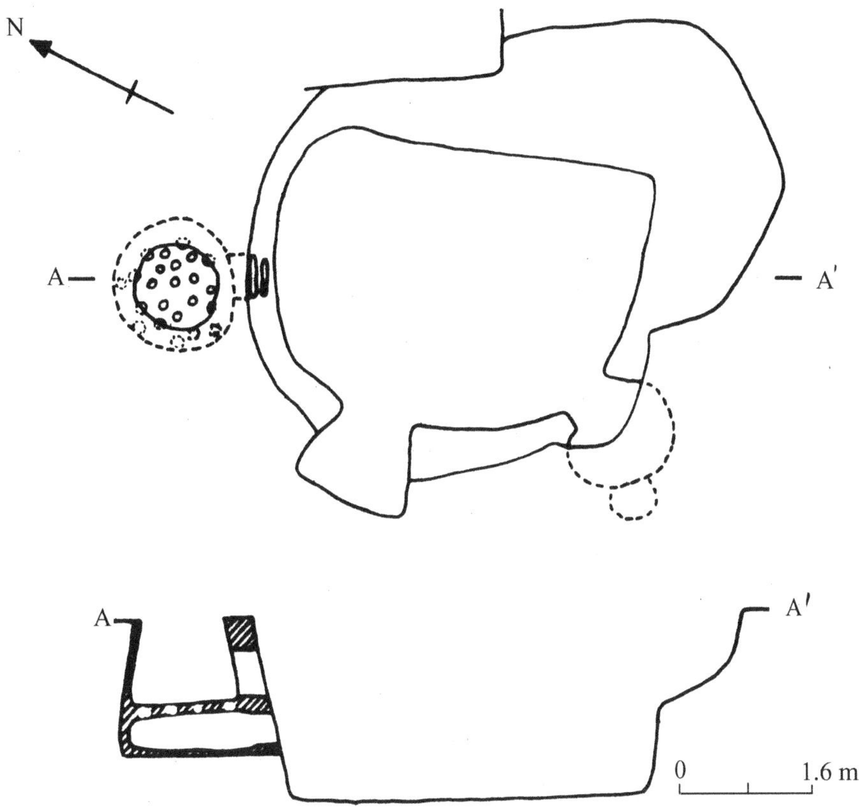

Figure 7-3. Plan and cross-section of the kiln Y9 at Nanguan (redrawn from the excavation record of Y9 kept in the Gucheng archaeological workstation).

burial was associated with the abandonment of the kiln and some kind of ritual ceremony.

Similar to the situation for the Miaodigou II remains at Dongguan, there are also many pits (especially bell-shaped pits) uncovered near the Upper Erligang kilns at Nanguan. Some of them probably functioned as the storage pits for pottery vessels.

A few Upper Erligang pottery-making tools, namely anvils and beaters have been found near some kilns (National Museum of Chinese History et al. 1996:221,226), so it is very likely that pottery manufacture was conducted nearby. Moreover, these kilns and their operating pits are very similar to those of the Miaodigou II period at the Dongguan site. It is reasonable to infer that these kilns, their working areas, nearby storage pits, and other facilities also constituted workshops, which facilitated the entire procedure of pottery manufacture, firing, and storage.

The eleven Upper Erligang kilns were dispersed over the southern part of the walled town, rather than concentrated in a small area (Fig. 7-4). In some circumstances, one workshop possibly had only one kiln. In other circumstances, one single workshop probably included two or more contiguous kilns, such as Y5 and Y6 in central position, and Y7, Y8, Y9 and Y10 in the western part of the southern town (see Fig. 7-4). According to the distribution of uncovered kilns, there were at least 5-6 workshops. Because most southern section of the town has not been excavated, we may expect that a lot of kilns beyond the excavated areas have not been discovered yet. Therefore, there were probably many more workshops consisting of single or multiple kilns at the southern part of the walled town. These individual workshops clustered in a limited area, forming a large-scale industrial complex. We may refer to this mode of production as "nucleated workshops", as defined by Peacock and Rice (see Chapter Four).

During the Erligang period, there are lots of sites where many kilns are distributed closely to form large-scale specialised areas of pottery production (nucleated workshops) (Yang 1992:490-493). Especially at the Zhengzhou site—the capital city of the Erligang state—there is a highly specialised centre of pottery production, where the distribution of kilns is similar to Nanguan (Henan Institute of Cultural Relics 2001). From archaeological evidence of other regions of the world, we can also find examples of nucleated workshops, such as in mid-third millennium B.C. northwest Mesopotamia, where early state societies also had developed (Wattenmaker 1998:51).

In contrast to an individual workshop industry, the essential characteristics of nucleated workshops are that pottery production became a major economic activity, cooperation in resource procurement or product distribution may exist in the industrial complex, production is usually year-round and conducted by full-time specialists, and products are highly standardised and of high quality (see Chapter Four). The last point will be demonstrated adequately by the analyses of indirect evidence of pottery production in this chapter.

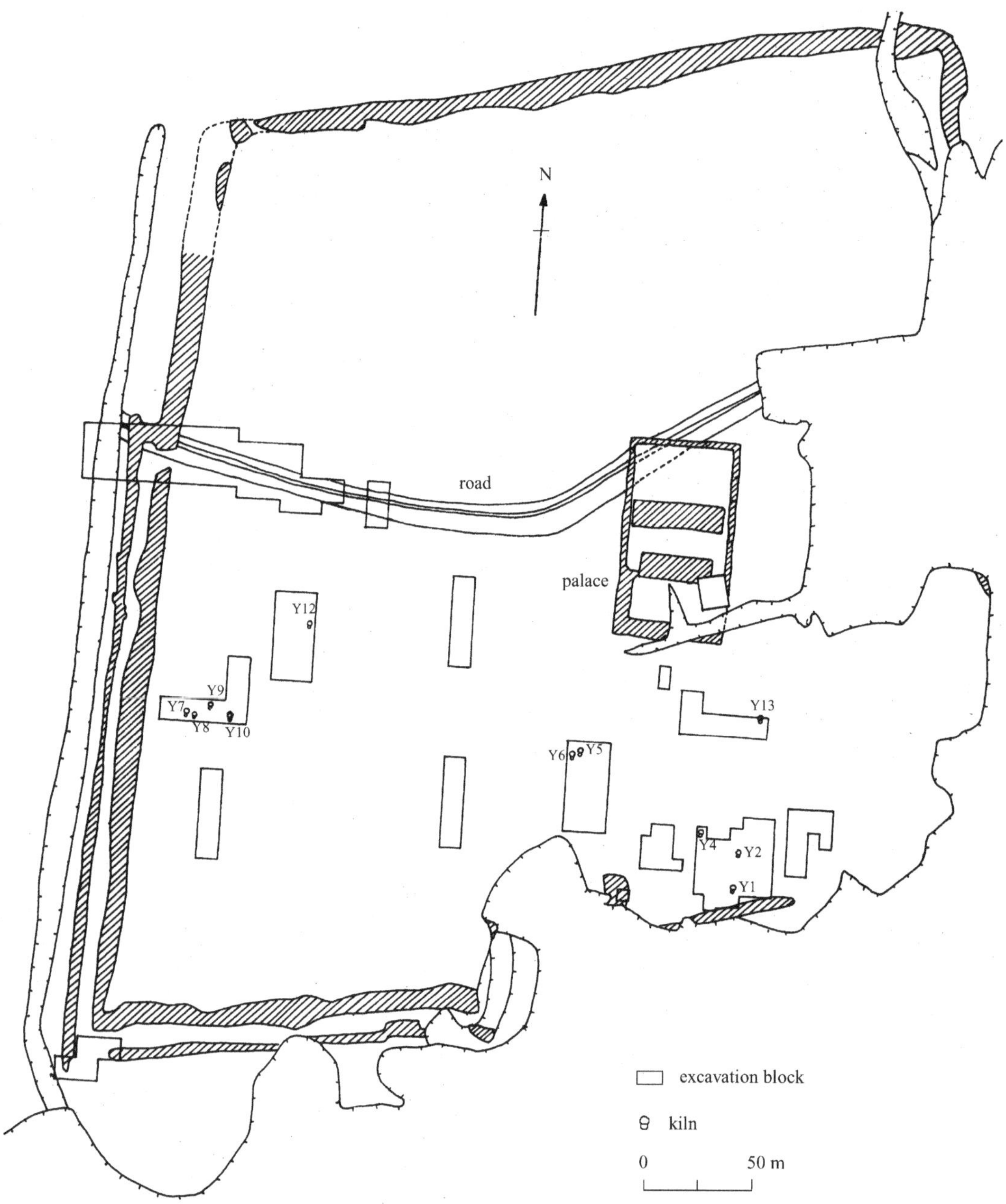

Figure 7-4. Map of the Nanguan site, showing the outline of the rammed-earth enclosure, positions of the palatial district and the Upper Erligang kilns (redrawn from the excavation record of the Nanguan site kept in the Gucheng archaeological workstation).

As discussed in Chapter Three, the walled town of Nanguan mainly functioned as a regional administrative centre, controlling the procurement and transport of copper and salt from the adjacent regions during the Erligang period. The most important building at Nanguan is the palatial compound located in the central portion of the walled town (see Fig. 7-4). It may have been the residence and administrative centre of the elite or officials dominating the town or even the whole basin. During the Upper Erligang phase, many pottery-making workshops were distributed at the south of the palatial district, and some of them were even in front of the palatial compound with a relatively close distance. This situation suggests that the Upper Erligang workshops for pottery production at Nanguan were very likely under the control of the elite or regional rulers.

However, it is not appropriate to simply define the pottery production in these workshops as "attached specialisation". Before identifying the social context of pottery production, we need to take into account the whole social background and productive structure of the Erligang period at Nanguan.

Only two Erligang buildings for commoners were found at Nanguan (one for Lower Erligang, and the other for Upper Erligang). It is possible that most houses either were situated outside the excavation areas, or were destroyed by later activities. Nonetheless, a large amount of evidence suggests that a community consisting of common people may have resided within the walled town, and the major residential area may have been located in the southeast section, where a great number of features and artefacts have been found. Besides pottery kilns, the majority of uncovered features are ash pits, which yielded a vast range of domestic refuse, including countless pottery sherds and animal bones, as well as various damaged ornaments, such as pottery loops, bone hairpins, and so on. Besides these objects relevant to ordinary life, there are also large quantities of artefacts for a variety of production activities, including: 1) agricultural tools, such as stone spades, stone axes, stone and pottery knives, and stone and pottery sickles (these tools can also be used for other activities); 2) hunting and fishing tools, or weapons, such as bronze, stone and bone arrowheads, bone awl and knives, and pottery net-sinkers; 3) weaving tools, such as pottery and stone spindle-whorls; 4) pottery-making tools, such as pottery beaters and anvils; and so forth (National Museum of Chinese History et al. 1996). These discoveries indicate that besides pottery production, the people living in this walled town were also engaged in diverse production activities, including agriculture, hunting and fishing, weaving, and so on. At least during the Upper Erligang phase, a number of kilns suggest that the pottery production became a large-scale and very important productive activity for this community.

The community in the Nanguan town was probably very different from other common communities or villages with respect to political and economic roles. The people of this community lived in a small walled town, only measuring 13.3 ha in area, and were adjacent to elite architecture (just in front of palatial district). It is inconceivable that this community was independent of elite control. Therefore, it is very likely that this community was directly under the control and administration of the elite or governmental officials, including various economic activities, such as agriculture, hunting and fishing, weaving, potting, and so on. The elite could obtain a variety of food, items and labour service through their control. Pottery production probably was an important economic activity controlled by the elite, especially during the Upper Erligang phase. In that case, pottery specialists may have been "attached" to the elite, rather than completely independent. A great majority of ceramics discovered at Nanguan are utilitarian goods, and there are only a very small number of high-value pottery wares, which did not occupy a salient position in social life, so the main function of specialised pottery production in the nucleated workshops must have been to produce utilitarian wares for the elite. The scale of this walled town and the palatial district (see Chapter Three) suggests that the number of elite members or governmental officials was very limited, and the community here was likely to be very small in population as well. It is possible that the large quantities of ceramics produced by so many workshops exceeded the needs of this town. If so, the major goal of pottery production here may have been to increase revenue for the elite, in addition to meeting everyday needs of the community. According to the current discoveries, Nanguan was possibly the only centre of pottery production during the Erligang period in the Yuanqu Basin. If so, it is possible that the elite at Nanguan monopolised the production and distribution of ceramics in the Basin, and thereby insured one of their revenue sources.

If the potters working in these Upper Erligang workshops were full-time retainers completely attached to the elite, the most possible pattern of workshops should be a segregated, highly specialised setting or facility, akin to "retainer workshop" defined by Costin (1991:9). However, as mentioned above, the distribution of Upper Erligang kilns or workshops at Nanguan was relatively dispersed and not aggregated together, and this nucleated workshop complex was actually composed of many dispersed individual workshops (see Figure 7-4), some of which probably adjoined residential areas or other activity areas (because a large quantity of ash pits and domestic refuse were found nearby). That is, there was not a segregated and separate workshop area for pottery production in this period at Nanguan. It is even possible that some workshops were combined with household structures, and we may call them "workshops in household settings". I therefore argue that the potters working in these workshops were not retainers completely attached to the elite, but rather communal members partially controlled and administered by the elite or governmental institutions. In that case, it is probable that the elite or governmental officials did not intervene each stage of productive process, but rather focused on controlling distribution of products. It is even possible that artisans could handle a part of their products. After completing the task of pottery production for the elite, those full-time potters, who probably kept certain independence, were also likely to participate in some other productive activities, especially the agricultural activities. Accordingly, the term "full-time" is relative rather than absolute, as discussed in Chapter Four.

In general, although the specialised pottery production during the Upper Erligang phase at Nanguan may have been controlled and administered by the elite or governmental institutions, the artisans working in these nucleated workshops were not retainers completely attached to elites. In other words, the degree to which pottery specialists were dependent on elites was possibly not very high. This case demonstrates the complexity and variability of productive context, and displays the possible relationship between producers and elite consumers beyond the dichotomy of "independent" and "attached" specialisation (see Chapter Four). Perhaps it is more feasible to refer to this kind of production as "elite-involved specialisation" rather than "attached specialisation".

An important phenomenon in the political economy of the Yuanqu Basin during the Erligang period is that some prestige goods relating to high-ranking status appeared. These prestige goods mainly include highly valued, relatively rare, and labour-intensive bronze vessels and jade items (Figure 7-5). These objects were found in a few high-ranking burials (National Museum of Chinese History et al. 1996:236-237,243-244; 1997a:19-21). Though some evidence shows that Nanguan was probably a crucial site in controlling and transporting copper for the Erligang state (see Chapter Three), there is no evidence of casting bronze vessels at Nanguan. During the Early Bronze Age of China, the production and distribution of the bronze vessels were basically monopolised by states (Liu and Chen 2003:133-134). Moreover, it is noteworthy that the bronze vessels found in Yuanqu are very similar in shape to their counterparts discovered in Zhengzhou—the capital city of the Erligang state (Henan Institute of Cultural Relics 2001:800, 804, 810). Likewise, there is no evidence of manufacturing jade objects in the Yuanqu Basin, and these jade objects also have the counterparts at Zhengzhou (Henan Institute of Cultural Relics 2001:684-685, 840-841). Therefore, it is very possible that the bronze vessels and jade items found in Yuanqu were obtained directly from the capital.

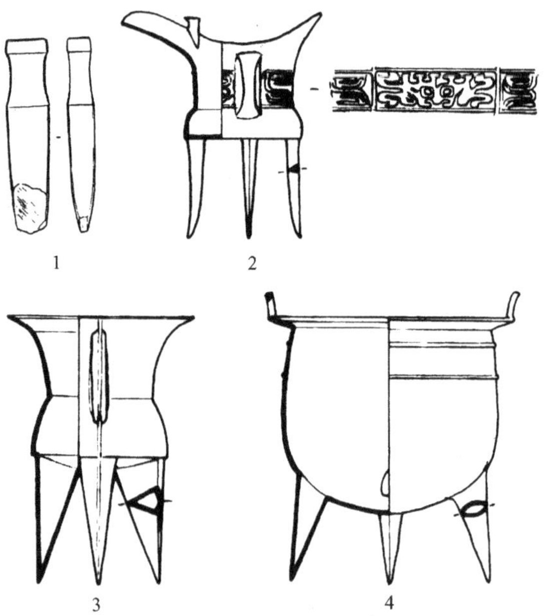

Figure 7-5. Jade and bronze objects of the Erligang period at Nanguan.
1. jade item (Lower Erligang); 2-4. bronze vessels (Upper Erligang), including *jue*, *jia*, and *ding* tripods.

The allocation of prestige goods could be used as a means of achieving vertical integration and maintaining a nested hierarchy of political leaders at the regional levels (Brumfiel and Earle 1987:3). In Erligang society, the central governments or royal elite groups could control local administrative centres more effectively by allocating some prestige goods to local rulers. Meanwhile, the local elite could legitimate and strengthen their dominant status and power, and sustain social inequality through monopolising these exotic prestige goods.

There are some valuable pottery vessels for the Erligang period at Nanguan. These ceramics can be categorised into three groups. 1) The elaborate ritual ceramic vessels that are usually found in graves, including *gui* stemmed dishes and *zunxingqi* jars (see Fig. 7-6: 1&4 and 7-7: 1&2). Some *gui* stemmed dishes are decorated with elaborate animal designs (see Fig. 7-7: 2), which are evidently duplicates of animal designs on some Erligang bronze vessels (see Henan Institute of Cultural Relics 2001:797-800, 820). 2) The ceramic copies of ritual bronze vessels. Currently only *jue* tripods have been discovered for the Erligang period at Nanguan (see Fig. 7-7: 3). These vessels were most likely used as drinking wares rather than ordinary serving items. 3) The proto-porcelain vessels. The proto-porcelain of the Erligang period is the earliest glazed porcelain in China, and it is relatively rare and valuable. The only vessel form of proto-porcelain recovered at Nanguan is the *cizun* jar (see Fig. 7: 4 & 5).

Proto-porcelain was made of kaolin. The paste of all proto-porcelain vessels of Nanguan is white and very hard, and most of them were glazed on the inner and outer surface and decorated with stamped checks. These vessels are completely different from other Erligang ceramics in shape, paste, colour and decoration. Furthermore, there are no records of kaolin resources in the Yuanqu Basin. Analyses of chemical composition also show that the chemical composition of proto-porcelain wares at Nanguan is different from that of other local pottery vessels, but similar to that of proto-porcelain wares discovered in southern China (Deng and Li 1996). In fact, based on neutron activation, all proto-porcelain wares discovered at several Erligang sites, including Zhengzhou (this analysis did not include Nanguan), were probably produced in southern China. Wucheng and its adjacent regions, in Jiangxi Province of southern China formed one major production centre (Chen et al. 1997; Chen et al. 1999, 2003). Therefore, the provenance of proto-porcelain of Nanguan may also have been in southern China (the Wucheng region) as well. In a word, the proto-porcelain vessels of Nanguan were imported rather than locally produced.

With respect to the source of the first two groups of valuable ceramics, there are two possibilities. One is that all of them came from the capital city like the bronze and jade objects, because they all had the equivalent wares at Zhengzhou (Henan Institute of Cultural Relics 2001:638, 644-645, 735, 741-747, 755). The other possibility is that they were locally produced, since there were no difficulties in raw materials and technology to duplicate these ritual ceramics in Yuanqu. It is regrettable that I could not conduct the neutron activation analysis for these vessels to identify their provenance.

Whether imported or locally made, these valuable ceramics are not the highest prestige goods. There is no evidence to show that they were exclusively associated with elite settings. They are even absent in elite burials which have bronze and jade items. Apart from some

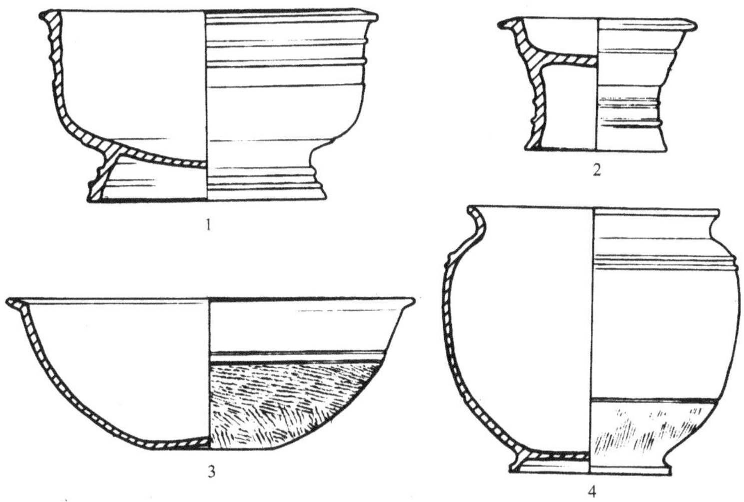

Figure 7-6. Ritual ceramic wares from the tomb M40 of Upper Erligang at Nanguan.
1. *gui* stemmed dish; 2. *dou* stemmed dish; 3. *pen* basin; 4. *zunxingqi* jar.

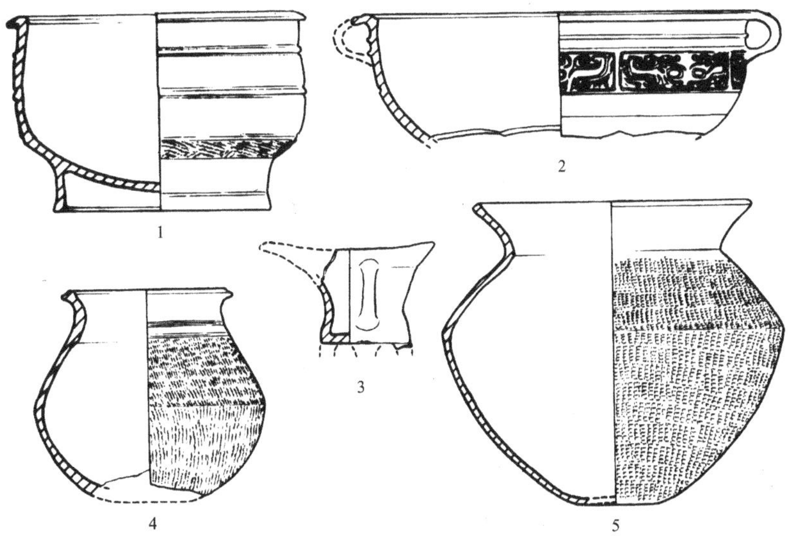

Figure 7-7. Valuable pottery vessels of the Erligang period at Nanguan.
1-2. *gui* stemmed dishes; 3. *jue* tripod; 4-5. *cizun* jars (1&4. Lower Erligang; 2, 3&5. Upper Erligang).

complete *gui* stemmed dishes and *zunxingqi* jars in several tombs, most sherds of these vessels are dispersed over an extensive area, including the ordinary ash pits near the kilns or in the residential areas. Therefore, it is likely these vessels were also used by individuals with relatively high status in the community, including the occupants of those tombs with ritual ceramics such as *gui* and *zunxingqi*. Namely, these valuable ceramics were not monopolised by the high elite, and there was no strict limitation for access to them. Meanwhile, these vessels occurred in very low frequency. Even if some of them were produced locally, the production of these valuable ceramics was unlikely to be the main goal and task for the pottery workshops at Nanguan. The primary objective of the specialised pottery production here was still probably utilitarian goods. This may be one of reasons that the elite or political leaders did not control and manipulate pottery production thoroughly, but rather only were "involved".

In the archaeological literature of other areas of the world, there are some cases where attached or "elite-involved" specialization produced utilitarian pottery goods (as mentioned in Chapter Four). In Late Uruk Mesopotamia, in addition to prestige goods, attached specialists also manufactured low-status goods including utilitarian ceramics (Stein 1996:32). In Early Dynastic Egypt, the production of diverse craft products, including utilitarian pottery was also controlled by the government (Wenke 1991:307-308). In the Inka Empire, some local potters were periodically recruited to produce Inka-style pottery in state-sponsored facilities. Similar to Yuanqu, the vast majority of these ceramics in Inka were extensively used as utilitarian goods. Only a small number of the finest vessels were more exclusively used by elites, but they were not the highest prestige goods. The specialists working in the Inka state-sponsored facilities were also local potters rather than retainers of the elite (Costin 1996). The case of Nanguan once again exhibits a particular mode of elite involvement in production of utilitarian pottery.

In ancient Chinese literature, we also can find some records that attached specialists produced both prestige and utilitarian goods for the elite in the Bronze Age. According to oracle-bone inscriptions, during the Shang dynasty, there existed *baigong*, meaning a number of kinds of craft specialists, and the government set up a special institution to control, administrate and sponsor the official craft economy. This sort of interrelationship between the government and craftspeople was later called *gongshangshiguan*, meaning that artisans and merchants were dependent on or attached to governments (Xu et al. 1994:632-633; Yang 1992:575-579). In the light of narrative of the *Zuo Zhuan*, a historical work from the Eastern Zhou period (771-221 B.C.), the Western Zhou (ca. 1046-771 B.C.) conquerors captured many Shang craft specialists, including potters, and forced them to serve the Zhou court (Yang 1981). This record suggests that like Western Zhou, the Shang royal elite group may have already controlled groups of "attached specialists", including some specialists producing utilitarian goods (Underhill 2002:234). In bronze inscriptions of the Western Zhou dynasty, there were many records of craft retainers associated with elite lineages (Zhu 2004:326-328). Attached or elite-involved specialisation therefore may have been a common characteristic for the craft production during the Bronze Age in China.

So far, I have discussed the mode of pottery production in the Erligang period, especially in the Upper Erligang phase at Nanguan, on the basis of direct evidence. For the Lower Erligang phase, there is insufficient direct evidence about pottery production, but it is still possible to make some reasonable inferences for its mode of production. Though only one pottery kiln has been discovered in this phase, I believe that there might be more kilns beyond the excavation areas in the walled town. Perhaps they are mainly distributed in northern portion of the site, where most features have been destroyed by later activities (see Chapter Three). Likewise, it is likely that these kilns and their workshops were also interrelated to the palatial district, and pottery production was performed in nucleated workshops controlled by the elite or governmental institutions. Above all, at Nanguan, the Lower and Upper Erligang phases had extremely similar assemblage constitutions in all kinds of artefacts, including bronze and jade objects, and ceramic vessels, and they were in the same social, political, and economic settings. Accordingly, we may safely presume that the pottery production during the Lower Erligang phase was also "elite-involved" specialisation, as an important component of elite political economy.

For the Late Erlitou period, no adequate direct evidence allows us to clearly outline the mode of pottery production at Nanguan, so we can only speculate based on the limited archaeological evidence. In this period, no kilns or other facilities of pottery manufacture have been found. The only evidence that pottery production happened here is from a few pottery-making tools (see the beginning of this section). Maybe the Late Erlitou kilns and other facilities were also situated somewhere beyond the excavated areas or have been destroyed by later activities, like those of Lower Erligang. It is hard to identify the organisation and scale of pottery production. However, the relatively high standardisation of Late Erlitou ceramics, which will be shown by indirect evidence, indicates that the pottery also was likely to be produced in specialised workshops by full-time artisans, similar to the Erligang period. Regarding the social context of pottery production, one possibility is that the pottery was produced by independent specialists during the Late Erlitou period at Nanguan, as at present no sufficient evidence indicates the existence of an elite group at Nanguan. Perhaps, elite-involved specialisation could appear in Fengcun, the largest and possible central place of the Erlitou period in the Yuanqu Basin (see Chapter Three). The other possibility is that elite-involved specialisation had occurred at Nanguan. The construction of the Nanguan Erlitou settlement was probably a strategic decision in relation to transporting important resources such as copper and salt (see Chapter Three), so Nanguan may have been an important settlement with political and strategic functions in the Erlitou period. The possibility that an elite group lived here and controlled craft production including pottery production cannot be excluded. In that case, it is likely that the elite structures and relevant facilities have not been found, or have been destroyed by later activities, like those contemporary kilns. Nevertheless, all these hypotheses need to be tested by archaeological discoveries in the future.

7.3. Pottery Classification and the Assemblage Composition

There are many common ceramic classes that are similar in shape attributes for the Erlitou and Erligang assemblages, but the general structure is different in vessel types and proportions of various ceramic classes between the two assemblages. There are even some variations in the assemblage compositions of ceramic classes between the Lower and Upper Erligang phase. I will investigate the variability in richness and evenness of ceramic types between three phases of the Nanguan site.

The excavations at the Nanguan site lasted from 1985 to 2003, and a large quantity of artefacts was discovered. Although a great number of pottery samples have been picked out, most sherds have not been systematically catalogued and they must still represent many pottery forms. The census of populations of ceramic classes cannot be carried out on the basis of current samples, which do not exactly represent the actual number and proportion of each ceramic class in the assemblage. In this study the classification and census of pottery types for the three phases are mainly based on the samples excavated in 1985 and 1986, which have been systematically processed and reported (National Museum of Chinese History et al. 1996). In addition, there are some vessel forms with a very few specimens (usually one or two) not included in the samples of 1985 and 1986. I add them into the pottery populations to complete the assemblages, and specify every type of them as "one" to indicate their rarity. Accordingly, the census of pottery populations for the three assemblages is not very precise but only an approximation.

I identify 34 ceramic classes of Late Erlitou from 242 specimens, 30 ceramic classes of Lower Erligang from 157 specimens, and 29 ceramic classes of Upper Erligang from 141 specimens (Fig. 7-8 ~7-11 and Appendix 6). Apparently, there is little variance for the richness of pottery types between Lower and Upper Erligang. Though the number of pottery types in the Erligang assemblage is slightly smaller than that in the Late Erlitou assemblage, the two assemblages are still approximate to each other in the richness of pottery types. Furthermore, compared with the Miaodigou II period, there is no significant variance in the richness of pottery classes during the Erlitou and Erligang periods. A number of vessel forms still reflect a variety of social requirements for diverse pottery functions.

Some significant variations had occurred in the distribution of pottery classes in the assemblage during the Late Erlitou and Erligang period at Nanguan (Fig. 7-11). For example, the top five ceramic types in Late Erlitou account for 58.27% of the total number, ranging from 3.72% to 19.42% (including the *zun* jar, *xiaokouguan* jar, *shoutiguan* jar, *yuanfuguan* jar, and *shenfuguan* jar). The top five ceramic types in the Lower Erligang phase account for 67.51% of the total, ranging from 9.55% to 25.48% (including the *li* tripod, *zun* jar, *xiaokouguan* jar, *shoutiguan* jar, and *gujianpen* basin). The top five ceramic types in the Upper Erligang phase account for 68.09% of the total, ranging from 6.38% to 31.21% (including *li* tripod, *xiaokouguan* jar, *shoutiguan* jar, *gujianpen* basin, and *dou* stemmed dish). Besides, all other pottery categories for the three assemblages are less than 5% of the total (see Fig. 7-11 and Appendix 6). Once again, the distribution of shape classes is very similar between the Lower and Upper Erligang phase, but it is different between the Late Erlitou and Erligang assemblage. From the Late Erlitou to Erligang period, the distribution of ceramic categories became increasingly concentrated to a few types, and the evenness of ceramic categories became increasingly low.

On the other hand, compared with the Miaodigou II assemblage (see Chapter Six), the distribution of ceramic classes in both Erlitou and Erligang assemblages became remarkably concentrated and fairly low in evenness. The pottery production during the Late Erlitou and Erligang periods began to concentrate on a few major pottery types. This change in diversity of assemblage composition of ceramic classes may imply a substantial increase in specialisation.

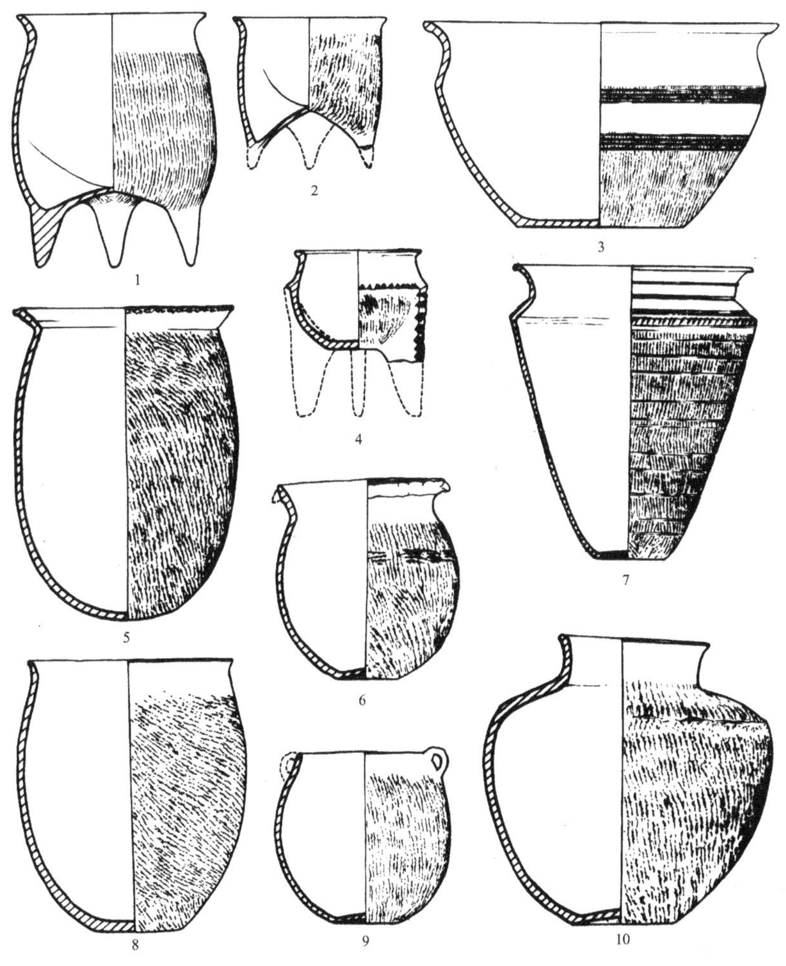

Figure 7-8a. Late Erlitou ceramics from the Nanguan site. 1. *dali* tripod; 2. *xiaoli* tripod; 3. *gujianpen* basin; 4. *ding* tripod; 5. *shoutiguan* jar; 6. *yuanfuguan* jar; 7. *zun* jar; 8.*shenfuguan* jar; 9. *bian'erguan* jar; 10. *xiaokouguan* jar.

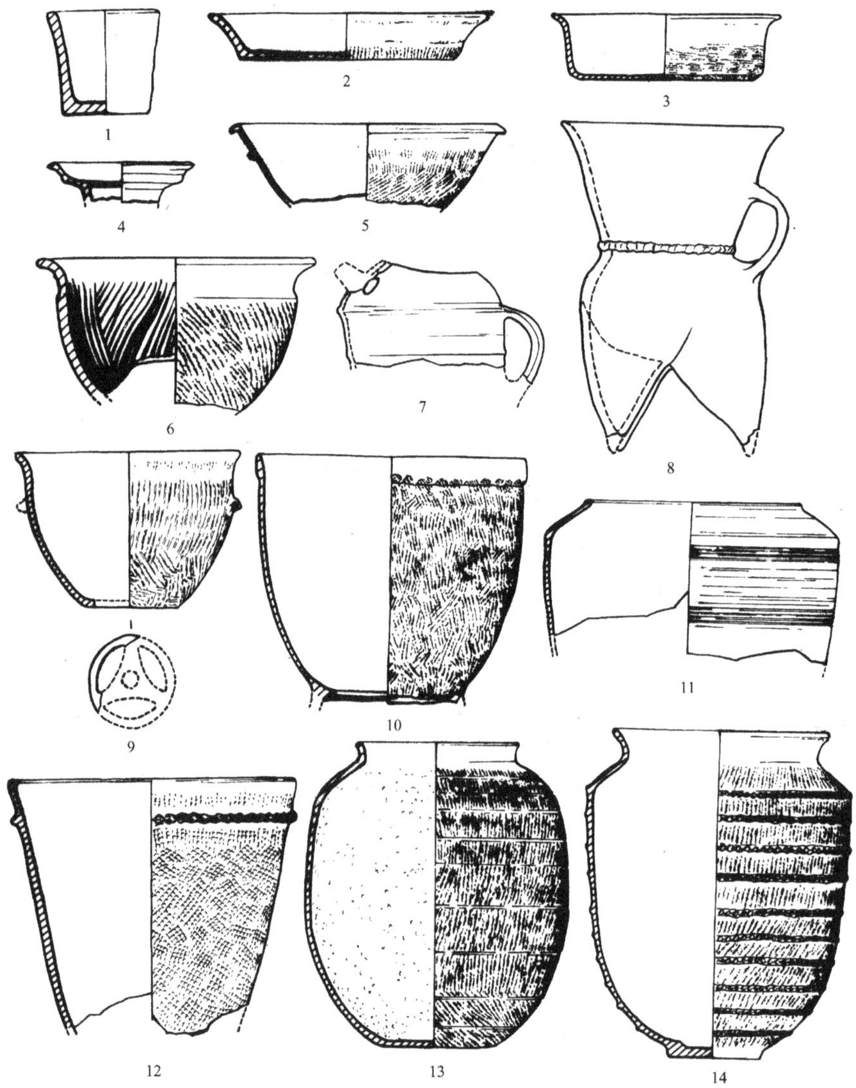

Figure 7-8b. Late Erlitou ceramics from the Nanguan site.
1. *bei* cup; 2. *pan* basin; 3. *zhifupen* basin; 4. *dou* stemmed dish; 5. *xiefupen* basin; 6. *kecaopen* basin; 7. *he* tripod; 8. *jia* tripod; 9. *zeng* steamer; 10. *yan* tripod; 11. *lei* jar; 12. *dakougang* jar; 13. *xiaokouweng* jar; 14. *dakouweng* jar.

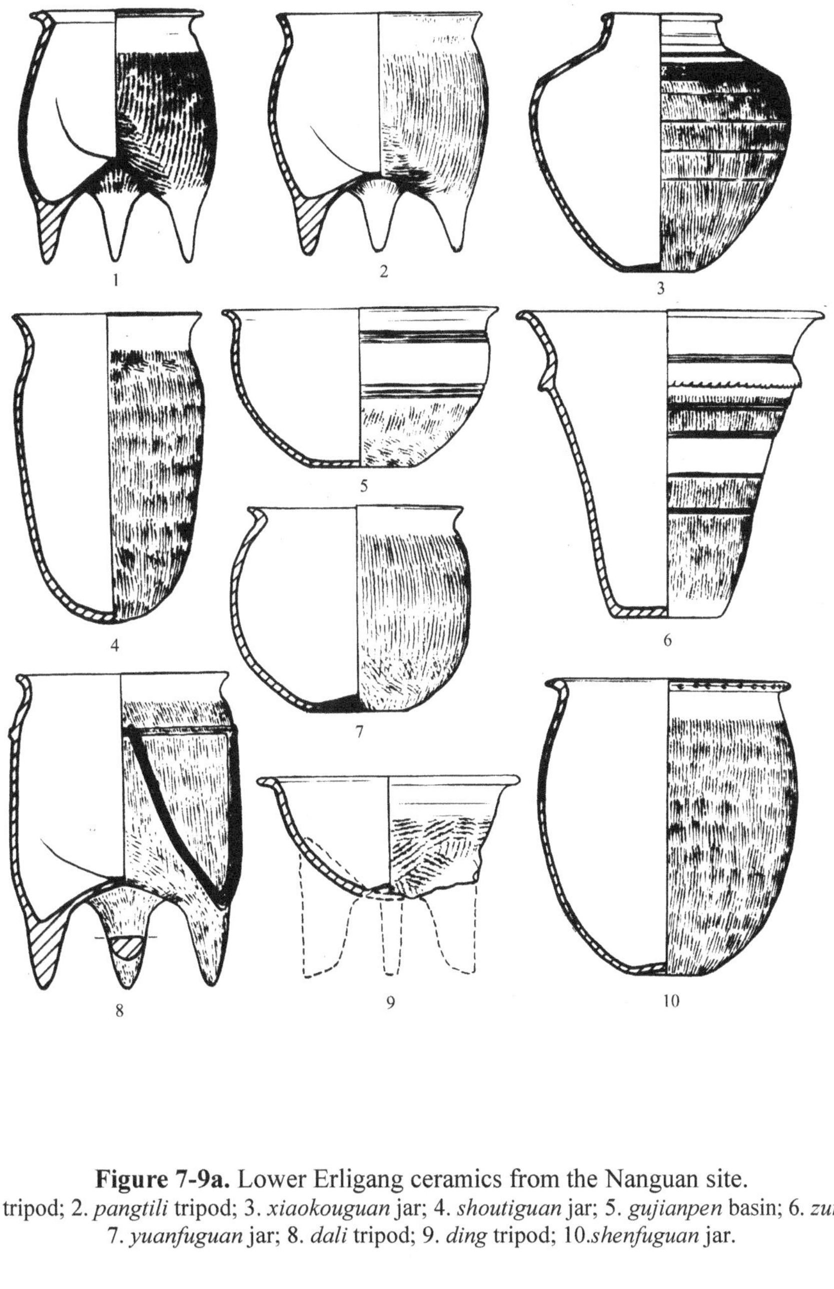

Figure 7-9a. Lower Erligang ceramics from the Nanguan site.
1. *li* tripod; 2. *pangtili* tripod; 3. *xiaokouguan* jar; 4. *shoutiguan* jar; 5. *gujianpen* basin; 6. *zun* jar; 7. *yuanfuguan* jar; 8. *dali* tripod; 9. *ding* tripod; 10.*shenfuguan* jar.

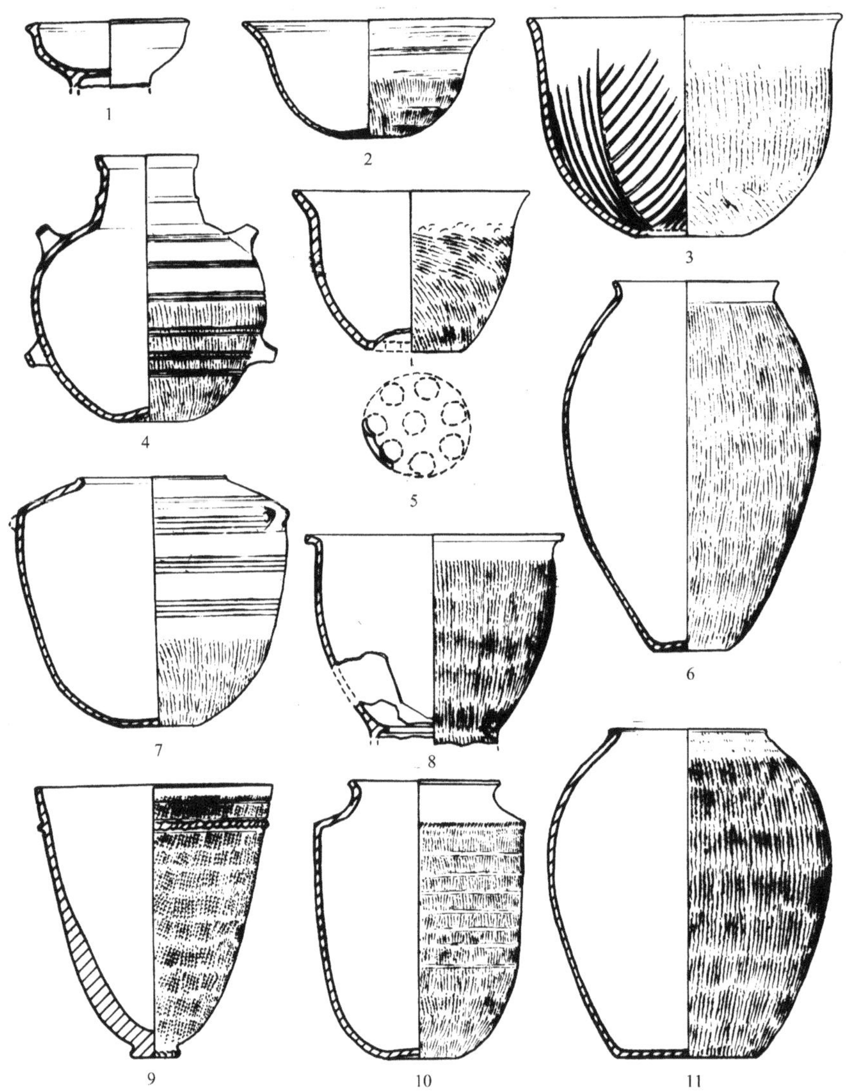

Figure 7-9b. Lower Erligang ceramics from the Nanguan site.
1. *dou* stemmed dish; 2. *xiefupen* basin; 3. *kecaopen* basin; 4. *si'erhu* bottle; 5. *zeng* steamer; 6. *gaotiguan* jar; 7. *lei* jar; 8. *yan* tripod; 9. *dakougang* jar; 10. *zhejianweng* jar; 11. *liankouweng* jar.

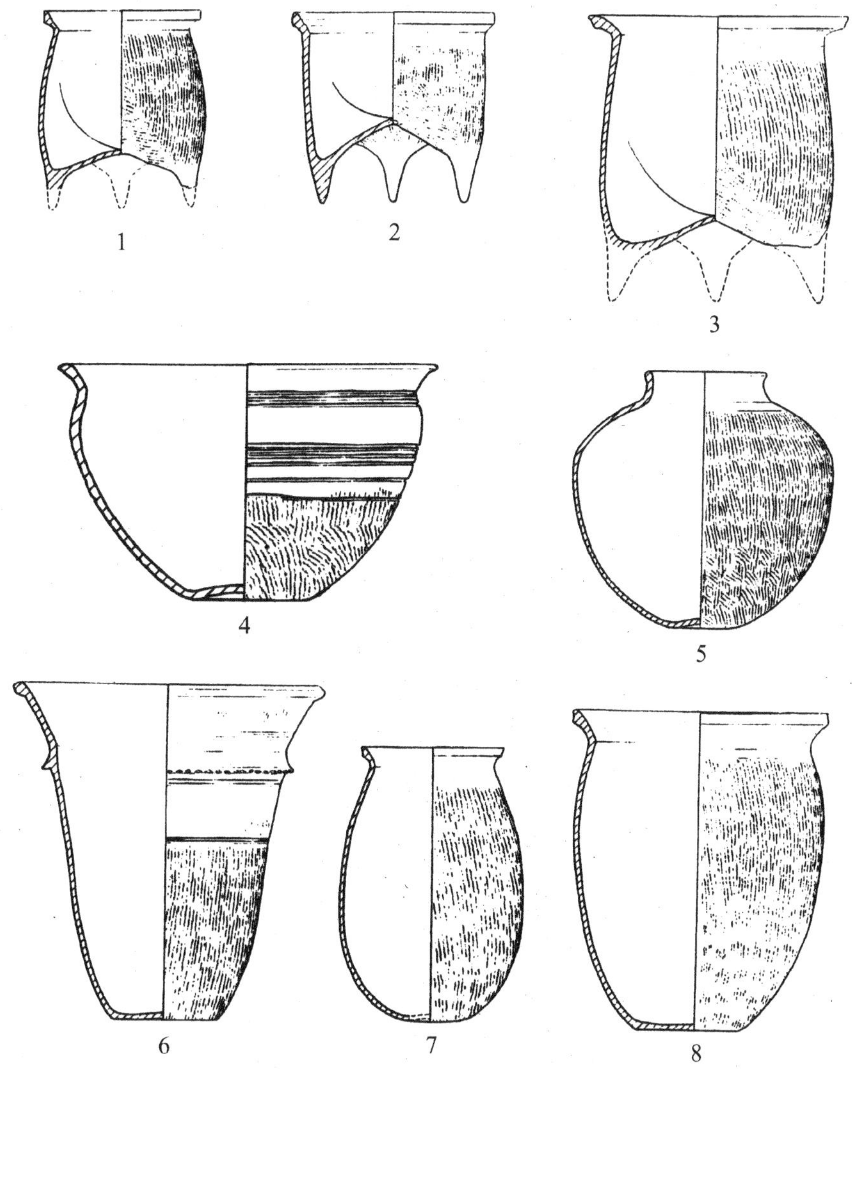

Figure 7-10a. Upper Erligang ceramics from the Nanguan site.
1. *li* tripod; 2. *liandangli* tripod; 3. *dali* tripod; 4. *gujianpen* basin; 5. *xiaokouguan* jar; 6. *zun* jar; 7. *shoutiguan* jar; 8. *shenfuguan* jar.

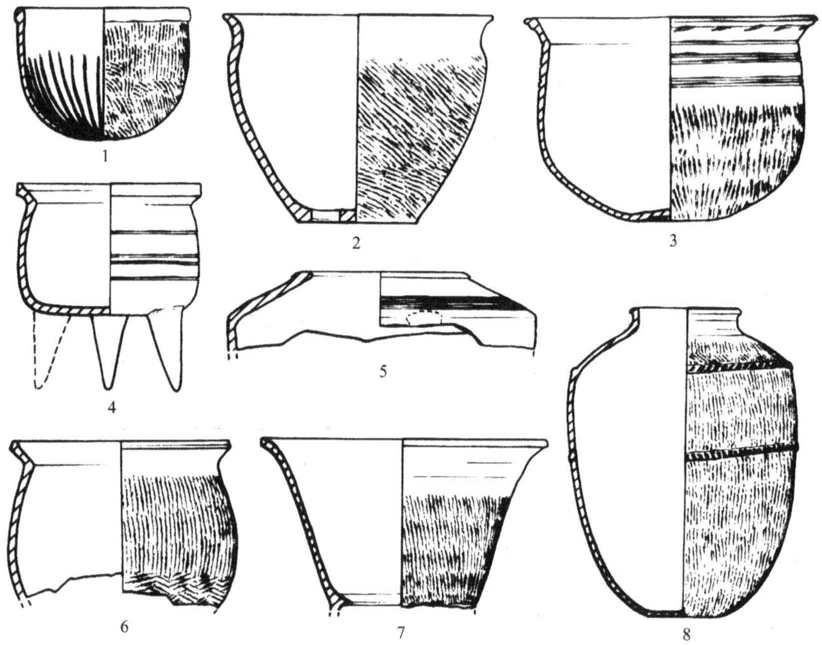

Figure 7-10b. Upper Erligang ceramics from the Nanguan site.
1. *kecaopen* basin; 2. *zeng* steamer; 3. *zhifupen* basin; 4. *ding* tripod; 5. *lei* jar; 6. *yuanfuguan* jar; 7. *yan* tripod; 8. *xiaokouwengg* jar.

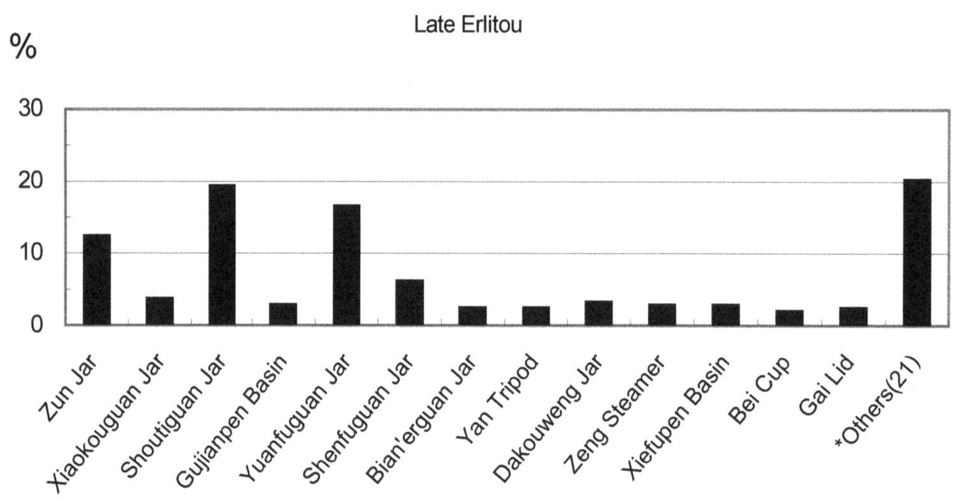

Figure 7-11a. Distribution of ceramic classes in the Late Erlitou assemblage.
*Others (21): 21 ceramic classes less than five specimens, including the *ding* tripod, *xiaoli* tripod, *dali* tripod, *zhejianguan* jar, *liankouguan* jar, *gaolingguan* jar, *bianfuguan* jar, *lei* jar, *dakougang* jar, *sanzuweng* jar, *xiaokouweng* jar, *wan* bowl, *shuangbanpen* basin, *kecaopen* basin, *zhifupen* basin, *shenfupen* basin, *pan* basin, *dou* stemmed dish, *jia* tripod, *he* tripod, and *jue* tripod.

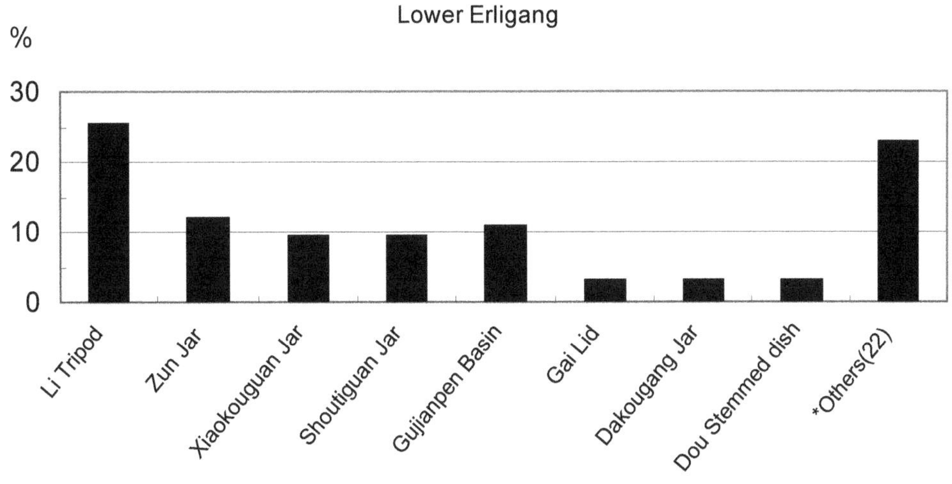

Figure 7-11b. Distribution of ceramic classes in the Lower Erligang assemblage.
*Others (22): 22 ceramic classes less than five specimens, including the *ding* tripod, *pangtili* tripod, *dali* tripod, *yan* tripod, *shenfuguan* jar, *yuanfuguan* jar, *dakouguan* jar, *gaotiguan* jar, *changjingguan* jar, *danxingweng* jar, *zhejianweng* jar, *liankouweng* jar, *lei* jar, *cizun* jar, *bo* bowl, *kecaopen* basin, *zhifupen* basin, *xiefupen* basin, *bei* cup, *si'erhu* bottle, *zeng* steamer, and *gui* stemmed dish.

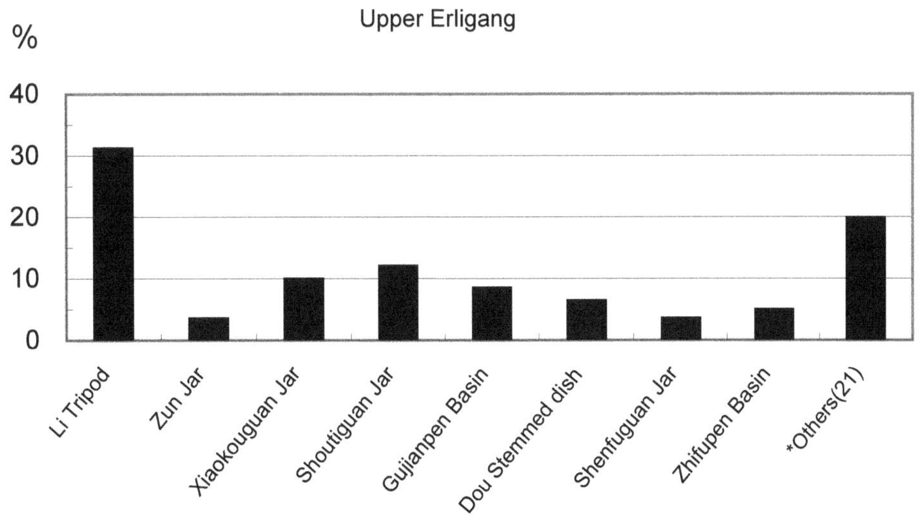

Figure 7-11c. Distribution of ceramic classes in the Upper Erligang assemblage.
*Others (21): 21 ceramic classes less than five specimens, including the *ding* tripod, *liandangli* tripod, *dali* tripod, *yan* tripod, *liankouguan* jar, *yuanfuguan* jar, *dakouguan* jar, *xiaokouweng* jar, *liankouweng* jar, *lei* jar, *cizun* jar, *bo* bowl, *kecaopen* basin, shuangban*pen* basin, *xiefupen* basin, *zhikoupen* basin, *bei* cup, *zeng* steamer, *jue* tripod, *gai* lid, and *gui* stemmed dish.

7.4. Indirect Evidence of Pottery Production

Compared with two previous periods, the nominal and metric variables during the Erlitou and Erligang periods changed qualitatively and showed different degrees of standardisation and diversity. I measured and recorded nominal and metric data of five major ceramic classes with relatively large sample size for statistical analyses, including the *li* tripod, *zun* jar, *xiaokouguan* jar, *shoutiguan* jar, and *gujianpen* basin (see Fig. 7-8~7-10). This includes almost all ceramic specimens of these five classes that were picked out from the collections between 1985 and 2003.

7.4.1. Analyses of Nominal Data

For the same reasons as that discussed in the previous chapter, the quantitative analysis of pottery colours in this chapter is still based on the investigation of sherds in the excavation report (Table 7-1) (National Museum of Chinese History et al. 1996:102, 167, 215). The general patterns of pottery colours in the Late Erlitou and Erligang assemblages are similar to those in the Miaodigou II assemblage. Besides a small number of red, reddish-brown, and black ceramics, a great majority of ceramics are grey and pale-brown. Especially in the Upper Erligang phase, the distribution of pottery colours was more concentrated on grey and pale-brown, so the diversity of pottery colours relatively decreased and the uniformity relatively increased. This change suggests that firing increasingly focused on the technology of reducing atmosphere. In addition, there were no significant correlations between pottery colours and shape classes.

With regard to pottery paste, similar to the Miaodigou II assemblage, sandy-paste vessels are still far more numerous than silt-paste vessels for the three phases of the Erlitou and Erligang periods (Table 7-2) (National Museum of Chinese History et al. 1996:102, 167, 215), and the number of sandy-paste vessels slightly increased from Late Erlitou to Upper Erligang. In addition, there is also a certain correlation between ceramic types and primary paste categories in the Late Erlitou and Erligang assemblages. For example, there is no or few silt paste for *li* tripods and *shoutiguan* jars, and they are basically composed of various sandy pastes. On the contrary, the paste texture is mainly silt paste for the *zun* jar, *xiaokouguan* jar and *gujianpen* basin, while diverse sandy-paste categories are generally small in number for each of them. On the other hand, those secondary sandy-paste categories do not strictly correlate to the pottery classes. Every ceramic type involves at least four paste categories (see Appendix 7).

Table 7-1. Frequency of pottery colours for the three phases of the Erlitou and Erligang periods at the Nanguan site.

	Grey and Pale-brown	Red and Reddish-brown	Black
Late Erlitou	83.5%	12.5%	4%
Lower Erligang	79.5%	14.3%	6.2%
Upper Erligang	92.6%	4.3%	3.1%

Table 7-2. Frequency of primary paste categories for the three phases of the Erlitou and Erligang periods at the Nanguan site.

	Sandy paste	Silt paste
Late Erlitou	64%	36%
Lower Erligang	68.8%	31.2%
Upper Erligang	70%	30%

However, different from the Miaodigou II assemblage, in the Late Erlitou and Erligang assemblages most ceramic classes focused on one major paste category, and the major composition of paste texture for a single ceramic type usually maintained coherent through time. During the whole Late Erlitou and Erligang period, for instance, the major paste category for the *li* tripod was medium sand; the major paste category for the *zun* jar, *xiaokouguan* jar, and *gujianpen* basin was silt (see Appendix 7). In short, during the Late Erlitou and Erligang periods, on the one hand, the paste texture displayed diversity as the past, on the other hand, the uniformity and standardisation in this variable were enhanced.

Uniformity and standardisation were exhibited more obviously in surface treatment and decoration of pottery during these two periods. According to the investigation on sherds in the excavation report (National Museum of Chinese History et al. 1996:102, 167, 215), among all surface treatments and decorations of the Late Erlitou and Erligang assemblages, cord-marking occupies an absolutely dominant position in number, followed by plain vessels, while basket-marking and check stamping are extremely rare. Other types of decoration, including clay strips, bowstrings, incised lines, and so on, are not usually used independently, but combined with cord-marking (Table 7-3). Although cord-marking is variable in width of marks and executive directions (vertical, slanting or horizontal) among individual vessels, all cord marks are executed with rolled-dowels. The high concentration of the surface treatment and decoration and the executive technology indicates the high degree of standardisation, which may have been related to highly specialised production.

The pottery forming techniques during the Late Erlitou and Erligang periods still included hand-building (coiling), moulding, and wheel-throwing. Moulding was mainly used for legs of *li* and *yan* tripods, and utilised in combination with coiling (Li 1996). In the Late Erlitou phase, a great majority of vessels were built by coiling, and moulding did not occupy a notable position, since the empty-leg tripods (*li* and *yan*) were very small in number in this phase (see Fig. 7-11a and Appendix 6a). During the Erligang period, including the Lower and Upper Erligang phases, moulding became a major building technique like coiling, because *li* tripods became a vessel form with the largest number of samples (see Fig. 7-11b & c, Appendix 6b & c). In contrast, wheel-throwing was still not popular during both Erlitou and Erligang periods,

Table 7-3. Frequency of surface treatments and decorations for the three phases of the Late Erlitou and Erligang periods at the Nanguan site.

	Cord-marking	Basket-marking	Check stamping	Plain	others
Late Erlitou	71.41%	0.7%	1.1%	16.6%	10.19%
Lower Erligang	72.03%	0.6%	0.3%	19.8%	7.27%
Upper Erligang	73.9%	0.37%	0.08%	21.5%	4.15%

and there were only a very small number of wheel-thrown vessels discovered at Nanguan. Beating, scraping, smoothing, and burnishing were still commonly used for shaping and finishing. Moreover, importantly, the slow-wheel technique was more extensively employed for pottery shaping during these two periods than during the Miaodigou II period. For the five ceramic classes I have investigated, in most situations the slow-wheel-shaped vessels are more than 50% of the total sample. In the Upper Erligang phase, the slow-wheel-shaped vessels are even more than 90% of the total sample for three ceramic types, including *li* tripods, *zun* jars, and *shoutiguan* jars (Appendix 8). The extensive utilisation of moulding and slow-wheel techniques was probably the important reason for dimensional standardisation, which will be sufficiently demonstrated in the following analyses of metric data.

In conclusion, though diversity of nominal variables still existed during the Late Erlitou and Erligang periods, standardisation and uniformity were increasingly present in every aspect, especially in the surface treatment and decoration, and manufacturing techniques. Compared with the Miaodigou II period, the specialised pottery production was doubtless further developed and intensified in these two periods.

7.4.2. Analyses of Metric Data

In this chapter, the analysis of variance (ANOVA) and investigation of standardisation of dimensions (by CV values) are conducted between two primary assemblages—Late Erlitou and Erligang, and two secondary assemblages—Lower Erligang and Upper Erligang. The statistical analyses for the two Erligang assemblages contain five major ceramic classes: the *li* tripod, *zun* jar, *xiaokouguan* jar, *shoutiguan* jar, and *gujianpen* basin. In the Late Erlitou assemblage of Nanguan, there are merely a few *li* tripods recovered, so the statistical analyses for this assemblage only address the other four ceramic classes, which are common vessel forms with relatively large sample size in the two primary assemblages.

As in the Miaodigou II period, there are more than 20 samples for most ceramic types in a specific chronological phase in the Erlitou and Erligang periods. However, many data sets for single variables are less than 20 in sample size, especially for vessel height, base diameter, and so forth, as most specimens are fragmental sherds rather than complete vessels. Here I follow the same principle as the former two chapters to keep or reject CV values according to the specimen number. Also, I will assess the significance of CV values less than 20 with those CV values more than 20 in sample size, especially with CV values of pooled data.

As showed in Chapter Four, I measured and recorded nine size variables altogether for all vessel forms (see Fig. 4-1). For the width of five marks, the analyses here only address cord-marking that occupies a dominant position among the decorations of the Late Erlitou and Erligang assemblages. There are strong correlations between some metric variables, such as between orifice diameter and neck diameter, vessel height and body height, and vessel height and leg length. Nonetheless, the correlation between orifice diameter and body width is not always strong, especially for *xiaokouguan* jars and *shoutiguan* jars in which there are not strong correlations between the two variables (Table 7-4). In addition, only *li* tripods address body height and leg length, which are not relevant to other four pottery types. Accordingly, as in the previous chapter, here I only offer the statistical data of seven metric variables: orifice diameter, body width (it refers to shoulder diameter for zun jars), base diameter, vessel height, rim length, width of five cord marks, and the ratio orifice diameter/vessel height (Table 7-5 and 7-6). Besides, a vast majority of *shoutiguan* jars are round-base vessels, so there is no base diameter available for this vessel form.

Table 7-4. Correlation coefficients among metric variables for the five ceramic classes in the Erlitou and Erligang assemblages of the Nanguan site.

		OD/ND*	OD/BW*	VH/BH*	VH/LL*
Li Tripod	Lower Erligang	0.866	0.859	0.873	0.826
	Upper Erligang	0.937	0.841	0.966	0.901
Zun Jar	Late Erlitou	0.962	0.781		
	Lower Erligang	0.862	0.630		
	Upper Erligang	0.891	0.802		
Xiaokouguan Jar	Late Erlitou	0.914	0.173		
	Lower Erligang	0.854	0.692		
	Upper Erligang	0.834	0.055		
Shoutiguan Jar	Late Erlitou	0.843	0.568		
	Lower Erligang	0.830	0.692		
	Upper Erligang	0.834	0.529		
Gujianpen Basin	Late Erlitou	0.962	0.934		
	Lower Erligang	0.948	0.831		
	Upper Erligang	0.743	0.734		

*OD/ND: Orifice diameter/Neck diameter. OD/BW: Orifice diameter/Body width.
VH/BH: Vessel height/Body height. VH/LL: Vessel height/Leg length.

Two inter-group changes of metric variables are of special interest. One is the variability between Late Erlitou and Lower Erligang, which represents the variance between the two primary assemblages. The other is the variability between Lower Erligang and Upper Erligang, which reflects the variance between the two secondary assemblages. I investigate these two variations through the analysis of variance (ANOVA). Before that, some data have to be rejected from this analysis due to small samples (less than six specimens for a data set) (Table 7-5).

Table 7-5a shows that there are no statistically significant variations for most variables of the four pottery classes between the Late Erlitou and Lower Erligang assemblages, and only a few variables of them vary significantly at the 0.05 level for ANOVA, such as the width of five cord marks for *zun* jars, the body width for *xiaokouguan* jars, and the orifice diameter and rim length for *shoutiguan* jars. Generally the variance of size variables between these two assemblages is very limited. In contrast, intriguingly, there are more significant variations of metric variables between the Lower and Upper Erligang phase (see Table 7-5b). Most variants of *li* tripods and *gujianpen* basins, and all available variants of *zun* jars vary significantly ($p<0.05$). Nevertheless, there are no or few significant variations for available variants of *xiaokouguan* jars and *shoutiguan* jars. In a word, some statistically significant variations for pottery dimensions had occurred between the Lower and Upper Erligang assemblages, but they are still limited and not very great.

Table 7-5a. Analysis of variance (ANOVA) for the four ceramic classes between the Late Erlitou and the Lower Erligang assemblages of the Nanguan site.

		Orifice diameter	Body width	Base diameter	Vessel height	Rim length	*Width/ 5M(CM)	*O/H
Zun Jar	F ratio	3.866	2.490	0.265	3.728	0.007	16.102	0.087
	P	0.054	0.120	0.611	0.067	0.936	<0.001	0.917
Xiaokouguan Jar	F ratio	0.203	23.689	--	--	1.458	0.815	--
	P	0.654	<0.001	--	--	0.233	0.373	--
Shoutiguan Jar	F ratio	8.651	0.557		--	5.072	0.495	--
	P	0.005	0.464		--	0.028	0.484	--
Gujianpen Basin	F ratio	1.267	0.182	--	--		1.276	--
	P	0.272	0.674	--	--		0.272	--

Table 7-5b. Analysis of variance (ANOVA) for the five ceramic classes between the Lower and Upper Erligang assemblages of the Nanguan site.

		Orifice diameter	Body width	Base diameter	Vessel height	Rim length	*Width/ 5M(CM)	*O/H
Li Tripod	F ratio	21.594	0.505		1.179	12.197	311.847	16.044
	P	<0.001	0.479		0.283	0.001	<0.001	<0.001
Zun Jar	F ratio	4.513	7.569	--	--	8.189	8.082	--
	P	0.038	0.008	--	--	0.006	0.007	--
Xiaokouguan Jar	F ratio	2.110	--	--	--	1.015	3.464	--
	P	0.151	--	--	--	0.317	0.068	--
Shoutiguan Jar	F ratio	2.352	1.020		--	0.000	30.496	--
	P	0.128	0.319		--	0.996	<0.001	--
Gujianpen Basin	F ratio	0.622	6.065	0.011	10.864		5.736	10.779
	P	0.438	0.022	0.918	0.006		0.027	0.007

Note: Significant at P<0.05.
*Width/5M(CM): Width/5 marks (Cord-marking). O/H: Orifice diameter/Vessel height.

The results of ANOVA suggest that the difference of metric data between the Late Erlitou and Erligang assemblage is very limited, and Table 7-6 also shows that the variable range of CV values for the Late Erlitou assemblage is very similar to that for the Erligang assemblage, so we may combine the two assemblages together for analyses and comparison.

As the Miaodigou II assemblage, there are still obvious differences for CV scores among distinct variants for those pottery types in the Late Erlitou and Erligang assemblage. CV values for rim length and width of five marks are remarkably higher than others; CV values for the ratio orifice diameter/vessel height are commonly lower than others; CV values for orifice diameter, body width, base diameter, and vessel height are relatively approximate to one another (Table 7-6). I will also analyse CV values of these distinct variables respectively.

Table 7-6. Sample means, standard deviations (SD), and coefficients of variation (C.V.) for the five classes of ceramics in the Late Erlitou and Erligang assemblages of the Nanguan site.

		Orifice diameter	Body width	Base diameter	Vessel height	Rim length	*Width/ 5M(CM)	*O/H
Li Tripod								
Lower Erligang	n=95	93	40		27	93	94	27
	Mean(cm)	15.58	17.61		21.30	2.51	1.43	0.72
	SD	1.75	1.98		3.04	0.37	0.28	0.07
	C.V.	**11.25%**	**11.22%**		**14.25%**	**14.71%**	**19.70%**	**9.24%**
Upper Erligang	n=153	143	42		25	153	128	25
	Mean(cm)	16.83	17.26		20.37	2.72	2.32	0.79
	SD	2.18	2.33		3.13	0.51	0.42	0.07
	C.V.	**12.96%**	**13.48%**		**15.34%**	**18.81%**	**18.19%**	**8.44%**
Combined	n=248	236	82		52	246	222	52
	Mean(cm)	16.34	17.43		20.85	2.64	1.94	0.75
	SD	2.11	2.16		3.08	0.47	0.57	0.08
	C.V.	**12.92%**	**12.37%**		**14.79%**	**17.94%**	**29.47%**	**10.06%**
Zun Jar								
Late Erlitou	n=39	31	27	12	10	31	30	10
	Mean(cm)	31.16	32.20	9.88	32.27	7.86	1.91	0.95
	SD	4.93	3.49	1.70	4.41	1.22	0.31	0.13
	C.V.	**15.83%**	**10.84%**	**17.16%**	**13.65%**	**15.53%**	**16.35%**	**13.20%**
Lower Erligang	n=48	36	35	16	14	44	35	14
	Mean(cm)	32.99	30.89	10.18	35.13	7.84	1.49	0.93
	SD	2.40	3.05	1.37	2.86	1.29	0.27	0.08
	C.V.	**7.28%**	**9.87%**	**13.42%**	**8.15%**	**16.52%**	**18.07%**	**8.73%**
Upper Erligang	n=32	24	17	5	4	28	13	4
	Mean(cm)	34.85	28.22			8.75	1.76	
	SD	4.37	3.73			1.36	0.37	
	C.V.	**12.54%**	**13.20%**			**15.52%**	**21.17%**	
Combined	n=119	91	79	33	28	103	78	28
	Mean(cm)	32.86	30.77	10.26	34.10	8.09	1.70	0.94
	SD	4.16	3.62	1.50	3.62	1.34	0.36	0.09
	C.V.	**12.66%**	**11.76%**	**14.61%**	**10.62%**	**16.56%**	**20.99%**	**10.09%**

		Orifice diameter	Body width	Base diameter	Vessel height	Rim length	*Width/ 5M(CM)	*O/H
Xiaokouguan Jar								
Late Erlitou	n=19	18	6	2	2	19	13	2
	Mean(cm)	13.71	27.15			2.99	1.75	
	SD	1.96	2.89			0.76	0.25	
	C.V.	**14.33%**	**10.64%**			**25.42%**	**14.16%**	
Lower Erligang	n=31	29	8	5	4	29	21	4
	Mean(cm)	14.37	33.29			3.28	1.62	
	SD	1.68	1.84			0.56	0.35	
	C.V.	**11.71%**	**5.53%**			**17.01%**	**21.34%**	
Upper Erligang	n=51	49	5	3	3	50	33	3
	Mean(cm)	13.82				3.48	1.94	
	SD	1.59				0.93	0.56	
	C.V.	**11.50%**				**26.78%**	**28.96%**	
Combined	n=101	96	19	10	9	98	67	9
	Mean(cm)	14.04	29.94	9.62	30.77	3.33	1.82	0.48
	SD	1.83	3.76	1.64	2.99	0.82	0.48	0.06
	C.V.	**13.05%**	**12.56%**	**17.08%**	**9.71%**	**24.55%**	**26.27%**	**11.84%**
Shoutiguan Jar								
Late Erlitou	n=31	31	12		4	31	28	4
	Mean(cm)	20.43	22.44			3.60	1.71	
	SD	1.80	1.86			1.02	0.45	
	C.V.	**8.81%**	**8.28%**			**28.30%**	**26.34%**	
Lower Erligang	n=35	34	12		3	34	35	3
	Mean(cm)	19.20	21.81			3.11	1.64	
	SD	1.58	2.28			0.70	0.35	
	C.V.	**8.21%**	**10.45%**			**22.50%**	**21.53%**	
Upper Erligang	n=78	74	26		8	77	71	8
	Mean(cm)	18.77	21.20		31.95	3.11	2.09	0.58
	SD	1.22	1.42		0.78	0.46	0.41	0.05
	C.V.	**6.52%**	**6.68%**		**2.44%**	**14.70%**	**19.77%**	**9.14%**
Combined	n=144	139	50		15	142	134	15
	Mean(cm)	19.25	21.64		32.15	3.22	1.89	0.62
	SD	1.59	1.80		1.94	0.70	0.46	0.07
	C.V.	**8.26%**	**8.31%**		**6.02%**	**21.71%**	**24.06%**	**11.04%**
Gujianpen Basin								
Late Erlitou	n=12	12	12	5	4		9	4
	Mean(cm)	28.18	26.75				1.68	
	SD	4.45	5.02				0.43	
	C.V.	**15.78%**	**18.77%**				**25.60%**	

		Orifice diameter	Body width	Base diameter	Vessel height	Rim length	*Width/ 5M(CM)	*O/H
Lower Erligang	n=16	14	12	8	7		13	7
	Mean(cm)	30.01	27.48	10.49	16.33		1.52	1.77
	SD	3.89	3.21	2.15	2.70		0.24	0.20
	C.V.	**12.96%**	**11.70%**	**20.45%**	**16.51%**		**16.09%**	**11.51%**
Upper Erligang	n=14	13	13	7	7		9	7
	Mean(cm)	31.12	30.62	10.59	21.83		1.89	1.46
	SD	3.37	3.16	1.28	3.50		0.49	0.16
	C.V.	**10.82%**	**10.31%**	**12.05%**	**16.02%**		**25.84%**	**10.74%**
Combined	n=42	39	37	20	18		31	18
	Mean(cm)	29.82	28.35	10.44	18.50		1.67	1.64
	SD	3.99	4.14	1.66	4.07		0.40	0.24
	C.V.	**13.39%**	**14.61%**	**15.88%**	**21.97%**		**24.06%**	**14.38%**

*Width/5M(CM): Width/5 marks (Cord-marking). O/H: Orifice diameter/Vessel height.

From Late Erlitou to Upper Erligang, the changes of CV scores for the five major ceramic classes do not display a common pattern. For *zun* jars and *xiaokouguan* jars, there are no identical varying trends among CV values of distinct variables. For *shoutiguan* jars and *gujianpen* basins, CV values for most but not all available variants tend to decrease over time, and this seems to indicate a certain increase of standardization. By contrast, CV values for most variables of *li* tripods become a little higher from Lower to Upper Erligang (Table 7-6). In short, although there are some changes for each ceramic type through time between the three phases, no one ceramic type exhibits a uniform and common changing trend for all metric variables, and there is certainly not a common changing tendency for all pottery classes. The variances of dimensional standardisation between the three assemblages are still limited. This conclusion is similar to the analytical result of ANOVA.

Now I set about to investigate the variances of CV scores for different size variables in the Late Erlitou and Erligang assemblages. For the five vessel classes between the three phases, the whole variable range of CV values for orifice diameter, body width, base diameter and vessel height is from 2.44% to 20.45%, and most of them range from 6% to 16%. The whole variable range of CV values for the ratio orifice diameter/vessel height is from 8.44% to 13.20%, and most of them range from 8% to 12%. The whole variable range of CV values for rim length is from 14.17% to 28.30%, and most of them vary between 14% and 26%. The whole variable range of CV values for width of five cord marks is from 14.16% to 28.96%, and most of them range from 16% to 26% (see Table 6-6). Remarkably, the variable ranges of CV scores for all metric variants between the three phases of the Late Erlitou and Erligang assemblages are clearly lower than those corresponding variable ranges in the Miaodigou II assemblage. Namely, most CV values of metric data in the Late Erlitou and Erligang assemblages are much lower than those in the Dongguan I and Miaodigou II assemblages (see Fig. 7-12).

The same situation also holds true for the pooled data of the two assemblages. The whole variable range of CV values for orifice diameter, body width, base diameter and vessel height is from 6.02% to 21.97%, and most of them vary between 8% and 15%. The whole variable range of CV values for the ratio orifice diameter/vessel height is from 10.06% to 14.38%, and most of them range from 10% to 12%. Likewise, the rim length and width of 5 marks have commonly higher CV values than other variables. CV scores for rim length range from 16.56% to 24.55%, and CV scores for width of five marks range from 20.99% to 29.47% (see Table 7-6). In general, the variations of CV values for combined data in distinct variables are similar to those between the three phases. On the other hand, these CV values of combined data in the Late Erlitou and Erligang assemblages are also much lower than those corresponding CV values in the Dongguan I and Miaodigou II assemblages (see Fig. 7-13).

Some CV values may have been influenced by small samples, but their impact is probably very limited. Apart from *gujianpen* basins that are less than 20 in sample size for each phase, most data sets of available metric variants (more than 5 individuals) of the other four ceramic classes are more than 20 in specimen number. Especially all metric variables of *li* tripods are more than 20 in number. Among all metric variables of these four ceramic types, the specimen number of orifice diameter and rim length is always more than or fairly approximate to 20. Moreover, a great majority of CV values of pooled data are more than or extremely approximate to 20 in sample size. Similar to Dongguan I and Miaodigou II, most CV scores for pooled data in the Late Erlitou and Erligang assemblages are still in the position between the lowest and highest values of the three phases (see Table 7-6). Compared with these data, most CV values whose samples are less than 20 do not exceed their variable range. Namely, most CV values with small samples may have not deviated far from the actual values, and they should be included in statistical analyses. Therefore, the standardisation index based on the metric data here may be reliable.

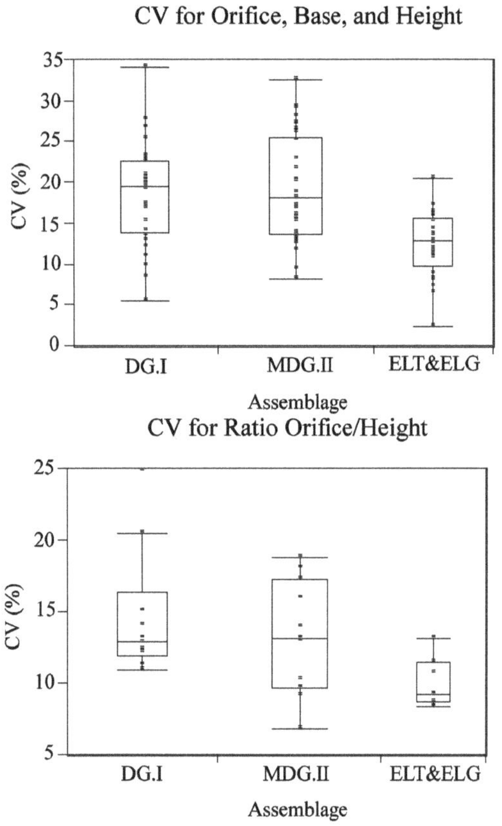

Figure 7-12. Box plots showing variable ranges of CV values for different chronological phases in Dongguan I, Miaodigou II and Erlitou & Erligang assemblages (DG..I=Dongguan I, MDG..II=Miaodigou II, ELT&ELG=Erlitou & Erligang).

Compared with the Miaodigou II assemblage, CV values for dimensional data in the Late Erlitou and Erligang assemblages generally become markedly low, indicating relatively high uniformity and standardisation. This change in standardisation index no doubt suggests a substantial increase in specialisation of pottery production.

In conclusion, two important results are derived from the analyses of metric data of Nanguan pottery. Firstly, although some significant variations had occurred for some variables of these major ceramic classes between the Late Erlitou and Erligang assemblage, and especially between the Lower and Upper Erligang phase, these variations are limited. These ceramic types do not display a uniform and common changing tendency between the three assemblages. The variability of dimensional standardisation between the three phases is also limited. This situation is similar to the Miaodigou II assemblage. Therefore, the pottery production from Late Erlitou to Upper Erligang may have been similar in productive intensity and the scale of organisation. Secondly, CV values of metric data for the three phases and combined data in the Late Erlitou and Erligang assemblages are commonly much lower than those corresponding CV values in the Dongguan I and Miaodigou II assemblages, indicating higher uniformity and standardisation. Undoubtedly, the specialised pottery production had been substantially developed in these two state societies. This conclusion is consistent with the analytical results of nominal data. These patterns in nominal and metric data suggest that during the Late Erlitou and Erligang periods, the pottery production may have been the large-scale and high-intensity activity. This analytical result supports the presumption in **Section 7.2** that potters of Late Erlitou and Erligang at Nanguan were probably "full-time" specialists.

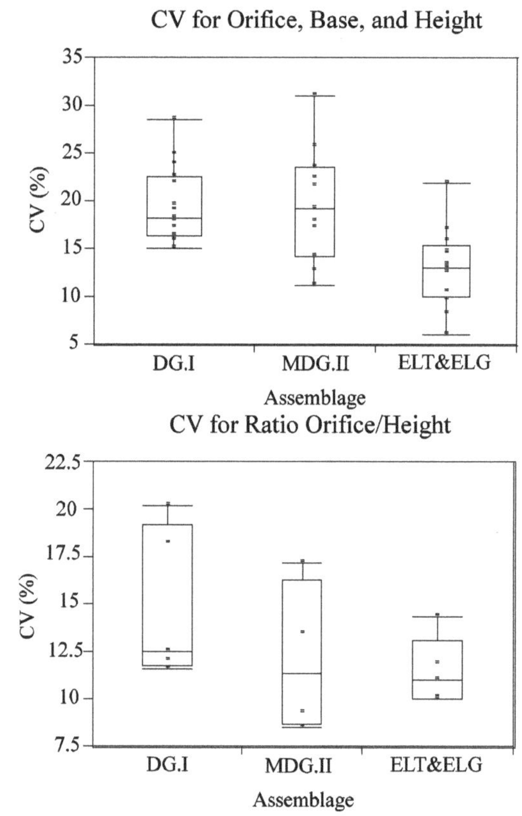

Figure 7-13. Box plots showing variable ranges of CV values for pooled data in Dongguan I, Miaodigou II and Erlitou & Erligang assemblages (DG..I=Dongguan I, MDG..II=Miaodigou II, ELT&ELG=Erlitou & Erligang).
Note: In this chart the Miaodigou II assemblage only includes the four major ceramic classes: *ding* tripods, *xiaoguan* jars, *daguan* jars, and *bo* bowls.

7.5. Summary

In this chapter, I have investigated the pottery production in a peripheral area of two state societies that developed successively. The discussions focus on the Nanguan site, which was the secondary regional centre in the Late Erlitou period and the primary central place in the Erligang period. During these two periods, the Yuanqu Basin witnessed a substantive change in productive mode, from the previous low-level specialisation to high-degree specialisation.

The surveys and excavations in the Yuanqu Basin show that pottery production was possibly concentrated on the regional level during the Erlitou and Erligang periods. This situation is very different from that in Dongguan I and Miaodigou II when production units may have been dispersed over the whole Basin. Especially in the Erligang period, Nanguan was perhaps the only production centre. This change reflected the development of craft specialisation, and related to the social and economic evolution in the Basin.

The direct and indirect evidence suggest that the pottery of Late Erlitou and Erligang discovered at Nanguan may have been produced in highly specialised workshops or workshop complex by full-time specialists. For the Upper Erligang phase, the direct evidence of pottery production is more adequate, from which I infer that the organisation of production for this phase probably was "nucleated workshops". Although the direct evidence for Late Erlitou and Lower Erligang is scanty, the indirect evidence in nominal and metric data of pottery sufficiently demonstrates that these two phases were possibly similar to Upper Erligang in the scale of organisation and the intensity of production. Both ANOVA and CV values of size variables suggest that the significant variations in pottery dimensions from Late Erlitou to Upper Erligang were very limited, so the ceramics of the three phases may have been the products of similar productive conditions. Compared with Dongguan I and Miaodigou II, the relatively high ceramic standardisation indicates the considerable development of specialised production during the Late Erlitou and Erligang periods.

The social context of pottery production for the Late Erlitou period at Nanguan is unclear. Based on current archaeological evidence, it is hard to judge whether the pottery production in this period at Nanguan was independent or attached/elite-involved specialisation, and each of these two possibilities cannot be rejected before new archaeological evidence is found. By contrast, the settlement configuration and the distribution of kilns (workshops) in the Erligang period suggest that the pottery production and distribution may have been controlled and manipulated by the elite or governmental institutions, and in the meantime it is possible that the potters were not retainers completely attached to the elite or political leaders, but rather communal members partially controlled by the elite. I thus define this mode of production as "elite-involved" specialisation, which mainly produced utilitarian wares rather than high-value items (actually there was not the highest prestige ceramics goods at Nanguan). Nanguan was perhaps the largest or the only centre of pottery production during the Erligang period in the Yuanqu Basin. By controlling or even monopolising the distribution of ceramic products, the elite or rulers could get a productive avenue to increase their revenue and wealth.

If attached or elite-involved specialisation had occurred during the Erlitou period, despite at Nanguan or Fengcun (the largest site), the pottery production in a common village, such as Guanjia (see the beginning of **Section 7.2**), may have been independent. Likewise, considering the limitation of archaeological discoveries, the possibility that during the Erligang period independent specialists produced pottery in a small village cannot be completely excluded either. If all these hypotheses are reasonable, it is likely that the "dual" craft economy had appeared in the Early Bronze Age in the Yuanqu Basin, like in the Late Uruk Mesopotamia: independent and some form of attached specialisation coexisted in a single society (Stein 1996).

During the Erlitou and Erligang periods, political economy in the early state societies had experienced dramatic changes in the Yuanqu Basin. The pursuit of social wealth by the elite may have directly led to the development of specialised pottery production. The intensification of production, including pottery production, was likely to be the result of changes in socio-political structure. The centralised control over political economy may have been the major cause that gave rise to highly concentrated and elite-involved specialisation for this period.

Chapter Eight

Conclusions

8.1. Introduction

This dissertation addresses two major issues: the development of social complexity and the changes of modes of pottery production from the Neolithic to early Bronze Age in the Yuanqu Basin, north-central China. Firstly I reconstruct the evolutionary process from tribe-level to state-level societies in the Basin, but the study in this aspect is not the emphasis of this thesis, and relevant analyses are relatively simple. On the basis of the socio-political backgrounds provided by the studies of settlement patterns, I focus on the changes of modes of pottery production and the development of specialisation in three distinct societies, which respectively represent the tribal-level, chiefdom-level, and state-level societies in the Basin.

In this last chapter, I intend to firstly summarise the analytical results of the previous chapters, and then further discuss some theoretical issues and assess some theoretical models in relation to pottery production, especially concerning the relationships between the mode of production and social complexity. Finally, I will point out some problems remaining in this study and some questions requiring further investigation in the future research.

8.2. Trajectories to Social Complexity

According to current archaeological data, the social evolution in the Yuanqu Basin may have experienced a process from tribe, chiefdom to state societies, or from local groups to regional polities (chiefdoms and states) (Johnson and Earle 2000).

During the Peiligang and Early Yangshao periods (ca. 6200-4100 B.C.), the social organizations in the Yuanqu Basin may have consisted of some autonomous local groups/tribal societies, which were likely to be organised on the basis of kinship, and there were no political hierarchies within and among communities. From the middle Yangshao to the Longshan Period (ca. 4100-2000/1900 B.C.), the Yuanqu Basin may have witnessed the development of complex societies, which were likely organised as distinct forms of chiefdom-level regional polities. Political hierarchy and social stratification commonly characterised these polities, but the population size, social scale and social integration appear to fluctuate periodically, and inter-group conflicts or warfare probably became increasingly intense through this long period of time. In the Early Bronze Age, or during the Erlitou and Erligang periods (ca. 2000/1900-1300 B.C.), the social entities in the Yuanqu Basin may have been integrated into two successive state systems—the Erlitou and Erligang states, and possibly played an essential role in procuring and transporting copper and salt resources distributed in southern Shanxi for the royal elite groups.

Some notions concerning social changes in this study are different from Railey's previous study. For instance, in the light of limited data, Railey denotes that the emergence of the complex society, namely the first political centralisation and settlement hierarchy, was likely to be in Late Yangshao rather than the Middle Yangshao period (Railey 1999:587-593).

Compared with other regions in the Middle Yellow River valley or the Central Plains, the process of social evolution in the Yuanqu Basin was probably similar to some regions, but different from others. Uneven development often occurred in different regions, and there was no a uniform model for socio-political changes for all regions in the Middle Yellow River. For example, the highly complex social system and relevant material remains during the Miaodigou II and Longshan periods in the Linfen Basin of southern Shanxi, have not been found at other areas, and the Taosi settlement system there was probably the most complex one in the Central Plains (Liu 2004:109-111, 170-176). Likewise, beyond the Central Plains, there may have been many different trajectories to the development of social complexity, especially in some regions in eastern China. Therefore, even though the archaeological data from China could demonstrate the evolutionary models from simple to complex societies, the processes and pathways of social evolution in different regions may have been very complex and variable. In brief, the abundant and distinctive archaeological records in China provide archaeologists with excellent data to reveal processes and dynamics of social evolution.

However, at present there is insufficient evidence in the Yuanqu Basin to illuminate the primary factors leading to the emergence and development of social complexity. As for the dynamics of the rise of early states in the Central Plains, this subject is beyond the research realm of this thesis, and it is impossible to solve this problem with the data of the Yuanqu Basin. However, it is worth noting that inter-group conflicts and warfare possibly played a crucial role for the development of social complexity, as well as for the origins of early states in the Central Plains.

8.3. Changes of Modes of Pottery Production and Development of Specialisation

This study has discussed the attributes and changes of modes of pottery production, and the development of specialisation in four periods (or three different forms of societies) in the Yuanqu Basin. The discussions were mainly based on four parameters offered by Costin

(1991): context, concentration, scale, and intensity of production. This analytical model for craft production has been productively employed by some archaeologists in other studies (e.g., Flad 2004). These four parameters were investigated through direct and indirect evidence of pottery production in this study.

The data for pottery analyses are mainly from the Dongguan and Nanguan sites in the Basin, which contain the archaeological complexes of Early Yangshao (Dongguan I), Miaodigou II, Late Erlitou and Erligang.

Although three of the four archaeological complexes (except Late Erlitou) can be divided into two or three chronological phases, the current data, especially the indirect evidence of pottery production suggest that there are few or very limited significant changes between the phases, and the mode of production may have retained coherent and stable within every complex. Hence I can concentrate on the comparisons between the four periods or complexes.

The direct and indirect evidence from the two sites sufficiently display the changes of modes of pottery production, and the development of specialisation in different social backgrounds, namely the tribal-level society (Early Yangshao), chiefdom-level society (Miaodigou II), and state-level society (Late Erlitou and Erligang).

The pottery production in the Yuanqu Basin experienced a process from dispersion to concentration at the regional level. During the Dongguan I (Early Yangshao) and Miaodigou II periods, pottery production may have been dispersed over many sites. For example, in the Miaodigou II period, four of five excavated sites have direct evidence of potting. During the Late Erlitou and Erligang periods, in constrast, direct evidence of potting is found in very few sites. Especially in the Erligang period, only one site, Nanguan, yielded remains linked to potting. Therefore, pottery production in these two periods may have been concentrated.

Correspondingly, the organisation, scale and intensity of pottery production within settlements also changed over time. In the Dongguan I period, there is insufficient evidence to show the emergence of pottery workshops at Dongguan and other contemporary sites in the Wei River valley. Pottery-making may have mainly taken place in domestic contexts. Meanwhile, the archaeological record suggests that not every household was engaged in pottery manufacture. Accordingly, it is very likely that the major organisation of production for this period was that of a "household industry". However, the possibility of workshops cannot be excluded completely, and the "individual workshop industry" was another possible organisation of production. Actually, household industry and individual workshop industry are similar in many aspects, and can be together defined as "household specialised production", in which pottery production is mainly carried out by part-time artisans. In contrast, during the Miaodigou II period, there is relatively definite evidence indicating the appearance of pottery workshops at Dongguan and Ningjiapo in the Yuanqu Basin. These workshops were either sparsely or relatively closely distributed in the villages, forming "individual workshop industry". Meantime, a household industry cannot yet be precluded completely from these settlements. Generally, the organisation of production in the Miaodigou II period probably sustained some similar traits like that of Dongguan I. The ceramic vessels in the two periods may have been made by part-time potters in small-scale productive units. This can be sufficiently demonstrated by the statistical analyses of dimensional data of pottery, which show relatively high CV values and low standardisation for metric variables for both periods. During the Late Erlitou & Erligang period, the modes of production may have changed substantively in the two state societies. Especially in the Upper Erligang phase, a number of individual workshops clustered in a limited area at Nanguan and composed a large-scale industrial complex, which could be regarded as "nucleated workshops". The relatively high standardisation indicated by both nominal and metric data suggests that ceramic production was possibly carried out by full-time specialists in these two periods.

Comparatively, the social context of pottery production may have experienced more distinctive changes between the three societies. During the Dongguan I period, at present there is not yet enough evidence at Dongguan to illustrate the context of production. One possibility is that producers were completely independent, and made ceramics for increasing their income. On the other hand, a few contemporary sites in the Wei River valley, Shaanxi Province, show that pottery kilns were frequently linked to different house groups (kin groups), while productive activities in the Early Yangshao societies may have usually been organised by diverse kin groups (Yan 1989a). Therefore it is possible the pottery production and distribution was at least partially controlled and managed by different kin groups, and potters were subsidiary to these kin groups on the basis of the division of labour. If so, this kind of productive context was also likely to exist in the Yuanqu Basin. During the Miaodigou II period, the uncovered kilns (workshops) at Dongguan and Ningjiapo are all located at edges of the villages, and no evidence indicates the linkage between the kilns/workshops and specific house groups. In the meantime, there were no prestige goods and elite structures associated with pottery-making facilities in these two sites. Accordingly, the pottery production may have been performed by independent craftspeople, rather than controlled and manipulated by kin groups or elite. Besides, the development of household roles and socio-economic hierarchies within communities during this period probably led to intensified competition among independent individual workshops or producers. During the Erlitou and Erligang periods, the development of socio-political structure in the state social systems may have given rise to the emergence of a new productive relationship. During the Erligang period, especially in the Upper Erligang phase, even if Nanguan was not the only place for pottery manufacture, it was most likely the largest productive centre in the Basin. It seems that those relatively dispersed kilns/workshops were associated with both the elite palatial district and the community of commoners, indicating that pottery specialists were neither independent nor completely attached to the elite, but rather partially controlled and administered by elite or governmental officials. One of the major objectives for pottery production at Nanguan may have been to increase revenue for the elite through the production of utilitarian pottery goods. This kind of mode of production may be defined as "elite-involved specialisation". For the Late Erlitou period, so far no adequate evidence indicates the social context of pottery production at Nanguan. Each of two possibilities that independent or attached/elite-involved specialisation existed here cannot be rejected before we find new archaeological evidence. Given that independent specialisation existed in a (or a few) small village beyond Nanguan or other central places (such as Fencun in the Erlitou period) for both periods, it is possible that the dual craft economy consisting of independent and attached/elite-involved specialisation had occurred during the Erlitou and Erligang periods in the Yuanqu Basin. This hypothesis needs to be tested by the future archaeological discoveries and studies.

Compared with modes of production, the development of specialisation from the Neolithic to the Early Bronze Age in the Yuanqu Basin is more adequately demonstrated in this study, especially by indirect evidence. Firstly, assemblage compositions of ceramic classes witnessed significant changes between the periods. From Early Yangshao to Miaodigou II and Erlitou & Erligang, the richness of ceramic types increased evidently, and this possibly related to the increase of social requirements for pottery use in complex societies. On the contrary, from Dongguan I and Miaodigou II to Late Erlitou and Erligang, the distribution of ceramic categories in the assemblages became increasingly concentrated on a few major types, showing increasingly low evenness of the distribution of ceramic classes and indicating the considerable improvement of specialised production. Secondly, pottery nominal variables (including pottery paste, surface treatment and decoration, manufacturing techniques, and colour) experienced a different fluctuation between the four periods. The relatively low diversity of nominal variables, as well as high correlation between shape classes and nominal variables in the Dongguan I assemblage perhaps stemmed from technological constraint, entrenched custom, or social control, rather than high specialisation. By contrast, the high diversity of nominal variants of pottery in the Miaodigou II assemblage probably related to many independent producers and less restriction to manufacture, and reflected lower specialisation. During the Late Erlitou and Erligang periods, standardisation and uniformity of nominal variables increased remarkably, indicating highly specialised pottery production. Thirdly, the development of specialisation is more clearly manifested by variations of CV (the coefficient of variation) values for metric variables of pottery over time. From Dongguan I to Miaodigou II, the variable ranges of CV scores for their common variants are similar to each other, and most CV scores are relatively high, indicating lower standardisation and specialisation. In the Late Erlitou and Erligang assemblages, CV values for vessel dimensions are commonly much lower than those corresponding CV values in the previous two assemblages, indicating higher-level standardisation of ceramic dimensions and specialisation of pottery production.

In brief, my analyses conform to Rice's prediction (Rice 1981), with an increase in diversity of pottery types and an increase in standardisation of ceramic shape and size over time during the developmental process of social complexity.

In general, during the Dongguan I and Miaodigou II periods, the degree of specialisation of pottery production was relatively low, but it became remarkably high in the Erlitou and Erligang periods, the two state-level societies. The Yuanqu Basin witnessed the substantial development of specialisation from the Neolithic to the early Bronze Age. This conclusion is completely different from Railey's. Based on very limited pottery data, Railey argues that ceramic production did not necessarily become an increasingly specialised pursuit over time, at least through the early Bronze Age, in the Yuanqu Basin and even ancient China (Railey 1999:613). Conversely, my study contends that the development of specialisation of pottery production was associated with social complexity, especially with the advent of early state societies in the Yuanqu Basin. As discussed above, during the Erligang period, due to the official control and elite involvement, the pottery production was concentrated in a large workshop complex, and may have been carried out by full-time artisans. Consequently, the ceramic products in this period became more standardised than those in the previous periods. These changes may have been related to increasingly centralised power in the area.

8.4. Productive Specialisation and Social Complexity

Based on these arguments, the development of craft specialisation or modes of production can be seen to positively correlate with the social evolution. This proposition supports Rice's theoretical model on the relationships between changes in the mode of production and the development of social complexity (Rice 1981).

However, there is no simple and definite correspondence between modes of production and social forms, and it is incorrect to simply view craft specialisation as a concomitant of social complexity. The discussion about their interrelations should be based on different degrees and/or types of specialisation. In this study, specialisation is viewed as any regularised production for consumers beyond household. As exemplified in this study, household specialised production (household industry or individual workshops) had appeared in tribal-level societies in the Early Yangshao period. Therefore, similar to some scholars (e.g., Clark and Parry 1990; Cross 1993; Underhill 2002),

I argue that specialisation is not necessarily linked only to complex societies; the initial specialisation might occur in a simple, non-stratified society, and may have been the result of extra-household division of labour within communities. At this point, different from Rice (1981; 1991) and some other scholars (i.e., Earle 1987), I do not think there is a necessary relation between the emergence of specialisation and the development of complex societies.

With the emergence and development of chiefdom-level complex societies, as exhibited by the Miaodigou II settlements in the Yuanqu Basin, perhaps the most marked change for the development of specialisation was the appearance and popularity of independent workshops (given there were no or few workshops in Early Yangshao). It appears that there were no substantive changes in the productive scale, intensity and the degree of specialisation. The workshops in the Yuanqu Basin are not a single case during the late Neolithic. A greater quantity of archaeological evidence of potting workshops has also been found in other regions of the Yellow River valley, such as Laohushan in Liangcheng, Inner Mongolia, and Baidaogouping in Lanzhou, Gansu Province (see Chapter Six).

In the Yuanqu Basin, in the early state societies, the specialisation of pottery production developed in many aspects. For example, pottery production may have mainly been conducted by full-time specialists. Above all, some kind of elite-involved specialisation characterised the production of this period. Elite-sponsored specialisation associated with elite political economy is the concomitant of complex societies. It could appear in chiefdom-level societies, but it is more frequently associated with state-level societies, as exemplified by the Yuanqu case.

Nonetheless, attached or elite-involved specialisation may not be the only mode of production in the early state societies of China. During the Erlitou and Erligang periods, attached/elite-involved specialisation may have mainly existed in the capital cities as Erlitou and Zhengzhou, and regional dominant centres like Nanguan. For most ordinary villages, independent specialists may have been major producers of ceramics. For example, it is possible the pottery production in the Nanguan and Guanjia (Henan) Erlitou settlements was independent (attached or elite-involved production perhaps existed at Fengcun, the central place in this period in the Basin). Likewise, if Nanguan was not the only site for pottery manufacture during the Erligang period, potters at other sites were likely to be independent specialists. This possibility of dual craft economy in state societies has been discussed in the preceding analyses.

With respect to causal factors leading to the origin and development of specialised pottery production, this issue has been discussed in Chapter Four. Similar to those suggestions, on the basis of the case studies in Yuanqu, I also ascribe relevant major causations to the following factors: supra-household divisions of labour (for the initial specialisation), population growth resulting in productive intensification and competition (for the development of specialisation), and elite control over political-economy for increasing revenue or monopolising prestige goods (for the emergence of attached or elite-involved specialisation). In addition, this study also supports the argument that attached (or elite-involved) specialisation is not a primary casual factor for the rise of social complexity and state formation, but rather a result (Earle 1987:67; Stein 1996:33; Underhill 2002:6). It appears that there is no attached (or elite-involved) craft specialisation until emergence of state-level societies in the Yuanqu Basin.

8.5. Questions for the Future Research

This study has outlined a framework on the social evolution, the change of modes of pottery production, and the development of specialisation in the Yuanqu Basin, and established a model for exploring relationships between productive modes, specialisation, and social complexity. However, due to a variety of reasons, especially the lack or scarcity of data, this study still leaves many unanswered or unverified problems, and raises many questions for the future research.

(1) Concerning the social evolutionary process in the Basin, this study addresses eight periods from the Neolithic to the early Bronze Age, but the reconstruction of social organisations or social structure for several periods is heavily dependent upon the macro-regional settlement patterns, and the data on the community level are often insufficient. Although seven sites in the Yuanqu Basin have been excavated, three of them have not been systematically reported (including Ningjiapo, Bancun and Guanjia), and this situation also brought about some difficulties for my analyses. In short, some suggestions about social patterns in this study are inferential rather than conclusive. The test of these propositions most relies on more archaeological reports and discoveries, especially more large-scale excavations at those large sites or central places.

(2) This study provides some valuable clues for illuminating the rise of complex societies and early states. For example, the emergence of the first complex society in the Yuanqu Basin was accompanied with the dramatic population growth, and the development of social complexity and the formation of state societies were closely associated with inter-group conflicts and warfare. Nevertheless, it is impossible to resolve the dynamics of the evolution of human societies only based on the current archaeological data in the Yuanqu Basin. The exploration of processes and dynamics of social complexity in ancient China will depend on a number of case studies at diverse regions, and will be a long-term task for Chinese archaeology.

(3) Due to the limitation of data, this study cannot involve each of eight periods for the reconstruction of productive modes. Though the four periods addressed in the analyses of pottery production represent three distinct social forms, the specification of productive modes for all periods can provide more sufficient insights into the

relationships between the development of modes of production and social evolution. This subject also has to be left to the future archaeological work and research.

(4) The reconstruction of modes of production is mainly based on direct evidence of pottery production. Though there are a number of pottery kilns and relevant facilities discovered at Dongguan and Nanguan, the direct evidence is not always adequate for each of the four periods involved in this study. For example, there are no excavated kilns and other productive facilities for Dongguan I; there are no any productive facilities of Late Erlitou discovered at Nanguan (only a few pottery-making tools were found). Therefore the recovery of productive organisations for these two periods can only be inferred in terms of relevant analyses and comparison, and the tests for these propositions have to depend on the future archaeological discoveries.

(5) The evaluation of standardisation and specialisation of pottery production is mainly based on indirect evidence. There are a great number of pottery specimens available for every assemblage or phase addressed in this study, but the number of samples is insufficient for some data sets of metric variables used for the statistical analyses. By comparison with those data sets sufficient in sample size, fortunately, it seems that most CV values with small samples do not deviate significantly from actual values, and they are still meaningful for statistical analyses. Nonetheless, larger samples will be more convincing.

(6) This study only investigates the pottery production in the Yuanqu Basin, but does not specially addresses pottery distribution. The current archaeological data do not allow us to discuss pottery distribution in the Basin in detail. In the future research, instrumental neutron activation analysis or X-ray spectroscopy analysis, and statistical analyses for nominal and metric variables of pottery among different contemporary sites may be viewed as useful approaches for investigating pottery provenience and distribution.

(7) In future research, the comparison between villages and central places, particularly between regional centres like Nanguan and capital cities like Zhengzhou, will enable us to understand craft production systems and political-economic structures of a society more profoundly.

(8) Interrelationships between variations of pottery assemblages and technological transformations or changes of ethnic groups are not systematically investigated in this study, and leave enough room for the future research.

8.6. Significance of this study

In spite of these unverified problems and unanswered questions, I still would like to see this study as a valuable contribution for understanding the relationships between craft production and social evolution in ancient China. Especially, the study on craft production is still a pioneering exploration for Chinese archaeology.

A long-standing topic of research in archaeology is the relationship between the development of complex societies in prehistory and changes in pottery production. Although many scholars working in different regions of the world have investigated this topic, few attempts have made with regard to China. This study adopts accepted methods used in some similar studies for investigating changes in settlement patterns and pottery production, reconstructs the developmental processes in social forms and modes of pottery production, and discusses the inter-relationships between them in a clearly defined geographical area in north-central China. This study will provide a comparable example for the studies in other areas of China.

Although in some cases I have to use small sample sizes, generally I have excellent samples of ceramics in comparison to some other studies. The large samples are very important for analyses of standardisation. Scholars who work in other regions of the world will find the valuable comparative data on pottery production from the Yuanqu Basin.

Based on the data from Yuanqu, I develop some new ideas on pottery production in prehistory. For example, "elite-involved production" in early state societies is a new concept suggested by this study.

Appendix 1. Neolithic and early Bronze Age sites in the Yuanqu Basin (PLG=Peiligang, EYS=Early Yangshao, MYS=Middle Yangshao, LYS=Late Yangshao, MGII=Miaodigou II, LS=Longshan, ELT=Erlitou, LELG=Lower Erligang, UELG=Upper Erligang, DG=Dongguan).

Site	Location	Size (ha)	PLG	EYS	MYS	LYS	MGII	LS	ELT	LELG	UELG
Tanjia	East bank of Yun River	3.5			3.5						
Chaijiagou	East bank of Yun River	1.92			1.92			1.92	1.92		
Yundong	East bank of Yun River	2.3		2.3 (DGII)		2.3					
Xiakou	East bank of Yun River	0.05				0.05					
Yuzi	East bank of Yun River	0.42							0.42		
Dongshi	East bank of Yun River	7							7		
Zhaozhai	Eastern Tableland	1.5					1.5	1.5			
Beibaie	Eastern Tableland	1.5						1.5			
Xiaozhao	North bank of Boqing River	15			15	8	8		8		
Shangbocheng	North bank of Boqing River	7.2			7.2		7.2	4.5			
Wangmao	North bank of Boqing River	1.5		1.5 (DGI)							
Xiguan	North bank of Boqing River	4.6							4.6	4.6	
Shijialing	North bank of Boqing River	1.5							1.5		
Beihe	North bank of Boqing River	2.25							2.25		
Xiwangmao	North bank of Boqing River	1.5							1.5		
Guanjia (Shanxi)	North bank of Yellow River	3.64			3.64	3.64	3.64	3.64	3.64		
zhaili	North bank of Yellow River	1.2			1.2	1.2	1.2	1.2	1.2		
Dongzhai	North bank of Yellow River	1			1		1		1		1
Nanguan	North bank of Yellow River	13.3				5.6			8.8	13.3	13.3

Site	Location	Size (ha)	PLG	EYS	MYS	LYS	MGII	LS	ELT	LELG	UELG
Xitan	North bank of Yellow River	0.5						0.5			
Ningjiapo	South bank of Boqing River	6		6 (DGI)	6		6		6	6	6
Xiabocheng	South bank of Boqing River	3.4			3.4		2				
Haoguduo	South bank of Boqing River	2			2			2			
Guanjia (Henan)	South bank of Yellow River	9	4.5		9				4.5		
Bancun	South bank of Yellow River	1.5	1.5		1.5		1.5				
Lingshang	South bank of Yellow River	1.5	0.5		1					1.5	
Rencun	South bank of Yellow River	6		2 (DGI)			4				
Yanghu	South bank of Yellow River	1				1					
Nancun	South bank of Yellow River	6						6			6
Xiwan	West bank of Jian River	0.5			0.5						
Hedi	West bank of Yellow River	2						2	2	2	
Baiquan	West bank of Yellow River	0.05								0.05	
Beibaotou	West bank of Yun River	30		4 (DGI)	30	8.4	12				
Diantou	West bank of Yun River	9			9						
Dongguan	West bank of Yun River	30		8 (DGI & II)	7.5	14	20	12			
Baigou	West bank of Yun River	3			3				3		
Nanbaotounan	West bank of Yun River	2			2				2	2	2
Xishi	West bank of Yun River	0.8		0.8 (DGII)							
Liujiazhuang	West bank of Yun River	0.56		0.56 (DGII)							
Nanbaotoubei	West bank of Yun River	0.5				0.5					

Site	Location	Size (ha)	PLG	EYS	MYS	LYS	MGII	LS	ELT	LELG	UELG
Xigou	West bank of Yun River	1.5							1.5		1.5
Beiyaozhuang	West bank of Yun River	0.75								0.75	
Shangdingban	West bank of Yun River	1.5					1.5				
Songcun	Western Tableland	0.8			0.8		0.8				
Fengcun	Western Tableland	30				2	10.5	30	20		
Beiyangbao	Western Tableland	2.5							2.5		
Total area (ha)			6.5	25.16	109.16	46.69	80.84	66.76	83.33	30.2	29.8
Number of sites			3	8	20	11	15	12	20	8	6

Appendix 2. The number and proportion of ceramic classes in the Dongguan I assemblage.

Ceramic classes	Number	%
Pingdibo Bowl	16	6.18
Gudibo Bowl	24	9.27
Gudipen Basin	45	17.37
Houdipen Basin	22	8.49
Qizuo Stand	38	14.67
Ping Bottle	16	6.18
Hu Bottle	5	1.93
Ding Tripod	6	2.32
Chikouguan Jar	21	8.11
Zheyanpen Basin	16	6.18
Zhaiyanpen Basin	7	2.70
Houdibo Bowl	11	4.25
Bei Cup	5	1.93
Gai Lid	7	2.70
*Others(8)	20	7.72
Total(22)	259	100

Appendix 3. The number and proportion of ceramic classes in the Miaodigou II assemblage.

3a. Early Miaodigou II.

Ceramic classes	Number	%
Penxingding Tripod	28	7.47
Xiaokouguan Jar	26	6.93
Xiaoguan Jar	41	10.93
Daguan Jar	28	7.47
Kecaopen Basin	14	3.73
Shuangbanpen Basin	17	4.53
Shuangbanbo Bowl	15	4.00
Bo Bowl	18	4.80
Jia Tripod	11	2.93
Fuzao Cook stove	12	3.20
Gang Jar	14	3.73
Aoxinpen Basin	33	8.80
Wan Bowl	21	5.60
Kuanyanpen Basin	9	2.40
Dou Stemmed dish	20	5.33
Gai Lid	6	1.60
Tongxingbei Cup	9	2.40
Changkoubei Cup	11	2.93
Quanzubei Cup	5	1.33
Others (14)	37	9.87
Total (33)	375	100

3b. Middle Miaodigou II.

Ceramic classes	Number	%
Guanxingding Tripod	28	11.48
Xiaokouguan Jar	13	5.33
Xiaoguan Jar	24	9.84
Daguan Jar	39	15.98
Kecaopen Basin	8	3.28
Shuangbanpen Basin	5	2.05
Shuangbanbo Bowl	8	3.28
Bo Bowl	14	5.74
Jia Tripod	16	6.56
Fuzao Cook stove	12	4.92
Wan Bowl	10	4.10
Kuanyanpen Basin	10	4.10
Gai Lid	11	4.51
Changkoubei Cup	8	3.28
Dan'erguan Jar	6	2.46
Sumianguan Jar	5	2.05
Others (11)	27	11.07
Total (27)	244	100

3c. Late Miaodigou II.

Ceramic classes	**Number**	**%**
Guanxingding Tripod	50	12.53
Xiaokouguan Jar	20	5.01
Xiaoguan Jar	26	6.52
Daguan Jar	30	7.52
Kecaopen Basin	9	2.26
Shuangbanpen Basin	11	2.76
Shuangbanbo Bowl	6	1.50
Bo Bowl	28	7.02
Jia Tripod	17	4.26
Fuzao Cook stove	33	8.27
Gang Jar	6	1.50
Wan Bowl	22	5.51
Kuanyanpen Basin	18	4.51
Dou Stemmed dish	5	1.25
Gai Lid	20	5.01
Tongxingbei Cup	13	3.26
Changkoubei Cup	7	1.75
Penxingding Tripod	9	2.26
Ping Bottle	7	1.75
Houdiwan Bowl	27	6.77
Dan'erbei Cup	9	2.26
Bizi Grate	7	1.75
Others (9)	19	4.76
Total (31)	399	100

Appendix 4. The compositions of ceramic paste for the five ceramic classes in the Miaodigou II assemblage.

Shape class	Total number	Silt		Medium sand		Coarse sand		Very coarse sand	
Ding Tripod		No.	%	No.	%	No.	%	No.	%
Early	28			2	7.14	13	46.43	13	46.43
Middle	28			15	53.57	8	28.57	5	17.86
Late	50			18	36.00	27	54.00	5	10.00
Combined	106			35	33.02	48	45.28	23	21.70
Xiaokouguan Jar									
Early	26	19	73.08	3	11.54	3	11.54	1	3.85
Middle	13	2	15.38	6	46.15	2	15.38	3	23.08
Late	20	12	60.00	2	10.00	6	30.00		
Combined	59	33	55.93	11	18.64	11	18.64	4	6.78
Xiaoguan Jar									
Early	41	1	2.44	14	34.15	18	43.90	8	19.51
Middle	24			3	12.50	10	41.67	11	45.83
Late	26			11	42.31	7	26.92	8	30.77
Combined	91	1	1.10	28	30.77	35	38.46	27	29.67
Daguan Jar									
Early	28			5	17.86	16	57.14	7	25.00
Middle	39			5	12.82	16	41.03	18	46.15
Late	30	2	6.67	6	20.00	17	56.67	5	16.67
Combined	97	2	2.06	16	16.49	49	50.52	30	30.93
Bo Bowl									
Early	18	13	72.22	4	22.22	1	5.56		
Middle	14	1	7.14	4	28.57	8	57.14	1	7.14
Late	28	8	28.57	14	50.00	5	17.86	1	3.57
Combined	60	22	36.67	22	36.67	14	23.33	2	3.33

Appendix 5. The Compositions of ceramic decorations for the eight ceramic classes in the Miaodigou II assemblage.

Shape class	Total number	Basket marks		Cord marks		Check marks		Plain	
Ding Tripod		No.	%	No.	%	No.	%	No.	%
Early	20	13	65.00	5	25.00	1	5.00	1	5.00
Middle	22	20	90.91	1	4.55	1	4.55		
Late	50	43	86.00	2	4.00	5	10.00		
Combined	92	76	82.61	8	8.70	7	7.61	1	1.09
Xiaokouguan Jar									
Early	18	18	100.00						
Middle	10	8	80.00	2	20.00				
Late	14	14	100.00						
Combined	42	40	95.00	2	5.00				
Xiaoguan Jar									
Early	39	23	58.97	14	35.90	2	5.13		
Middle	24	10	41.67	13	54.17			1	4.17
Late	25	8	32.00	17	68.00				
Combined	88	41	46.59	44	50.00	2	2.27	1	1.14
Daguan Jar									
Early	27	13	48.15	13	48.15			1	3.70
Middle	37	7	18.92	26	70.27	4	10.81		
Late	30	17	56.67	12	40.00	1	3.33		
Combined	94	37	39.36	51	54.26	5	5.32	1	1.06
Bo Bowl									
Early	18	9	50.00					9	50.00
Middle	14	7	50.00	1	7.14			6	42.86
Late	28	8	28.57	5	17.86	1	3.57	14	50.00
Combined	60	24	40.00	6	10.00	1	1.67	29	48.33
Kecaopen Basin									
Early	14	9	64.29	5	35.71				
Middle	8	7	87.50					1	12.50
Late	9	6	66.67	3	33.33				
Combined	31	22	70.96	8	25.81			1	3.23
Shuangban pen Basin									
Early	17	13	76.47	1	5.88			3	17.65
Middle	5	5	100.00						
Late	11	10	90.91	1	9.09				
Combined	33	28	84.85	2	6.06			3	9.09
Shuangbanbo Bowl									
Early	15	9	60.00	1	6.67			5	33.33
Middle	8	3	37.50					5	62.50
Late	6	4	66.67					2	33.33
Combined	29	16	55.17	1	3.45			12	41.38

Appendix 6. The number and proportion of ceramic classes in the Late Erlitou and Erligang assemblages.

6a. Late Erlitou

Ceramic classes	Number	%
Zun Jar	30	12.40
Xiaokouguan Jar	9	3.72
Shoutiguan Jar	47	19.42
Gujianpen Basin	7	2.89
Yuanfuguan Jar	40	16.53
Shenfuguan Jar	15	6.20
Bian'erguan Jar	6	2.48
Yan Tripod	6	2.48
Dakouweng Jar	8	3.31
Zeng Steamer	7	2.89
Xiefupen Basin	7	2.89
Bei Cup	5	2.07
Gai Lid	6	2.48
Others (21)	49	20.25
Total (34)	242	100

6b. Lower Erligang

Ceramic classes	Number	%
Li Tripod	40	25.48
Zun Jar	19	12.10
Xiaokouguan Jar	15	9.55
Shoutiguan Jar	15	9.55
Gujianpen Basin	17	10.83
Gai Lid	5	3.18
Dakougang Jar	5	3.18
Dou Stemmed dish	5	3.18
Others (22)	36	22.93
Total (30)	157	100

6c. Upper Erligang

Ceramic classes	Number	%
Li Tripod	44	31.21
Zun Jar	5	3.55
Xiaokouguan Jar	14	9.93
Shoutiguan Jar	17	12.06
Gujianpen Basin	12	8.51
Dou Stemmed dish	9	6.38
Shenfuguan Jar	5	3.55
Zhifupen Basin	7	4.96
Others (21)	28	19.86
Total (29)	141	100

Appendix 7. The compositions of ceramic paste for the five ceramic classes in the Late Erlitou and Erligang assemblages (ELT = Erlitou, LEG = Lower Erligang, UEG = Upper Erligang).

Shape class	Total number	Silt		Fine sand		Medium sand		Coarse sand		Very coarse sand	
Li Tripod		No.	%	No.	%	No.	%	No.	%	No.	%
ELT	4										
LEG	95			6	6.32	68	71.58	17	17.89	4	4.21
UEG	153			8	5.23	78	50.98	50	32.68	17	11.11
Combined	248			14	5.65	146	58.87	67	27.02	21	8.47
Zun Jar											
ELT	39	29	74.36			1	2.56	6	15.38	3	7.69
LEG	48	37	77.08			3	6.25	5	10.42	3	6.25
UEG	32	24	75.00			5	15.63	3	9.38		
Combined	119	90	75.63			9	7.56	14	11.76	6	5.04
Xiaokou guan Jar											
ELT	19	16	80.00							4	20.00
LEG	31	21	67.74	1	3.23	4	12.90	3	9.68	2	6.45
UEG	51	25	49.02	5	9.80	15	29.41	4	7.84	2	3.92
Combined	101	62	60.78	6	5.88	19	18.63	7	6.86	8	7.84
Shoutiguan Jar											
ELT	31					4	12.90	16	51.61	11	35.48
LEG	35			4	11.43	12	34.29	12	34.29	7	20.00
UEG	78	1	1.28	4	5.13	20	25.64	28	35.90	25	32.05
Combined	144	1	0.69	8	5.56	36	25.00	56	38.89	43	29.86
Gujianpen Basin											
ELT	12	8	66.67					2	16.67	2	16.67
LEG	16	11	68.75			1	6.25	3	18.75	1	6.25
UEG	14	9	64.28			1	7.14	2	14.29	2	14.29
Combined	42	28	66.67			2	4.76	7	16.67	5	11.90

Appendix 8. The number and proportion of pottery vessels by slow-wheel shaping for the five ceramic classes in the Late Erlitou and Erligang assemblages (ELT = Erlitou, LEG = Lower Erligang, UEG = Upper Erligang).

Shape classes	Total number	Number of slow-wheel shaping	%
Li Tripod			
ELT	4		
LEG	93	66	70.97
UEG	153	150	98.04
Combined	246	216	87.80
Zun Jar			
ELT	36	27	75
LEG	46	30	65.22
UEG	31	30	96.77
Combined	113	87	76.99
Xiaokouguan Jar			
ELT	19	14	70
LEG	30	18	60
UEG	51	43	84.31
Combined	100	75	75.00
Shoutiguan Jar			
ELT	27	4	12.90
LEG	34	12	35.29
UEG	78	71	91.03
Combined	139	87	62.59
Gujianpen Basin			
ELT	12	7	58.33
LEG	16	9	56.25
UEG	13	7	53.85
Combined	41	23	56.10

Bibliography

Adams, R. E. W. and Richard C. Jones, 1981, Spatial Patterns and Regional Growth among Classic Maya Cities. *American antiquity* 46(2): 301-322.

Allan, Sarah, 1984, The Myth of the Xia Dynasty. *Journal of the Royal Asiatic Society of Great Britain and Ireland* 2: 242-256.

1991, *The Shape of the Turtle: Myth, Art and Cosmos in Early China*. Albany, State University of New York.

Allen, Kathleen M. S., 1992, Iroquois Ceramic Producrion: A Case Study of Household -Level Organization. In *Ceranic Production and Distribution: An Integrated Approach*, edited by George J. Bey III and Christopher A. Pool. Boulder, Westview Press: 133-154.

Archaeological Team Leader Training Program, 1999, Zhengzhou Xishan Yangshao Shidia Chengzhi De Fajue (Excavation of Walled Town of the Yangshao Period at Xishan, Zhengzhou). *Wenwu* 7: 4-15.

Arnold, Dean E., Hector A. Neff, Ronald L. Bishop and Michael D. Glascoc, 1999, Testing Interpretative Assumptions of Neutron Activation Analysis: Contemporary Pottery in Yucatan , 1964-1994. In *Material Meaning: Critical Approaches to the Interpretation of Material Culture*, edited by Elizabeth S. Chilton. Salt Lake City, The University of Utah Press: 61-84.

Arnold, Dean E. and Alvaro L. Nieves, 1992, Factors Affecting Ceramic Standardization. In *Ceramic Production and Distribution: An Integrated Approach*, edited by George J. Bey III and Christopher A. Pool. Boulder, Westview Press: 93-113.

Arnold, Jeanne E. and Ann Munns, 1994, Independent or Attached Specialization: The Organization of Shell Bead Production in California. *Journal of field archaeology* 21 (4): 473-489.

Arnold, Philip J., III, 1991, *Domestic Ceramic Production and Spatial Organization: A Mexican Case Study in Ethnoarchaeology*. Cambridge, Cambridge University Press.

Bagley, Robert, 1999, Shang Archaeology. In *The Cambridge History of Ancient China*, edited by Michael Loewe and Edward Shaughnessy. Cambridge, Cambridge University Press: 124-231.

Banpo Museum, Shaanxi Archaeological Institute and Lintong County Museum, 1988, *Jiangzhai*. Beijing, Wenwu Press.

Beijing University, 2003, *Huaxian Quanhucun (a Report on the Quanhu Site, Huaxian)*. Beijing, Kexue Press.

Beijing University, Hubei Institute of Archaeology and Jingzhou District Museum, 1992, Shijiahe Yizhi Diaocha Baogao (an Investigation Report on Shijiahe Sites). *Nanfang Minzukaogu* 5: 213-294.

Beijing University and Nanyang Institute of Cultural Relics in Henan, 1998, Henan Dengzhou Baligang Yizhi Fajue Jianbao (Preliminary Report on the Excavatins at the Baligang Site, Dengzhou, Henan). *Wenwu* 9: 31-45.

2000, Henan Dengzhou Baligang Yizhi 1998 Niandu Fajue Jianbao (Brief Report of 1998 Excavation Season at the Baligang Site, Dengzhou, Henan). *Wenwu* 11: 23-31.

Benco, Nancy L., 1988, Morphological Standardization: An Approach to the Study of Craft Specialization. In *A Pot for All Reasons: Ceramic Ecology Revisited*, edited by Charles C. Kolb and Louana M. Lackey. Philadelphia, PA, Temple University: 57-72.

Blackman, M. James, Gil. J. Stein and Pamela B. Vandiver, 1993, The Standardization Hypothesis and Ceramic Mass Production : Thchnological, Compositional , and Metric Indexes of Craft Specialization at Tell Leilan, Syria. *American antiquity* 58(1): 60-80.

Blanton, Richard E., Gary M. Feinman, Stephen A. Kowalewski and Linda M. Nicholas, 1999, *Ancient Oaxaca: The Monte Alban State*. Cambridge, Cambridge University Press.

Blanton, Richard E., Stephen A. Kowalewski, Gary M. Feinman and Jaura M. Finsten, 1993, *Ancient Mesoamerica: A Comparison of Change in Three Regions*. New York, Cambridge University Press.

Blanton, Richard E., S. Kowalewski, G. Feinman and J. Appel, 1981, *Ancient Mesoamerica*. Cambridge, Cambridge University Press.

1982, *Monte Alban's Hinterland. Part 1: The Prehispanic Settlement Patterns of the Central and Southern Parts of the Valley of Oaxaca, Mexico, Memoir No.15*. Ann Arbor, University of Michigan Museum of Anthropology.

Blanton, Richard, Gary Feinman, Stephen Kowalewski and Peter Peregrine, 1996, A Dual-Processual Theory for the Evolution of Mesoamerican Civilization. *Current anthropology* 37(1): 1-14.

Brumfiel, Elizabeth M., 1998, The Multiple Identities of Aztec Craft Specialists. In *Craft and Social Identity*, edited by Cathy Lynne Costin and Rita P. Wright. Arlington, Virginia, the American Anthropological Association. No. 8: 145-152.

Brumfiel, Elizabeth M. and Timothy K. Earle, 1987, Specialization, Exchange, and Complex Societies: An Introduction. In *Specialization, Exchange, and Complex Societies*, edited by Elizabeth Brumfiel and Timothy Earle. Cambridge, Cambridge University Press: 1-9.

Carneiro, Robert L., 1970, A Theory of the Origin of the State. *Science* 169: 733-738.

1981, The Chiefdom: Precursor of the State. In *The Transition to Statehood in the New World*, edited by Grant D. Jones and Robert R. Kautz. Cambridge, Cambridge University Press: 37-75.

Chang, Kwang-chih, 1968, Toward a Science of Prehistoric Society. In *Settlement Archaeology*, edited by Kwang-chih Chang. Palo Alto, California, National Press: 1-9.

1986, *Archaeology of Ancient China*. New Haven, Yale University Press.

Chen, Tiemei, G. Jr. Rapp, Zhichun Jing and Nu He, 1997, Zhongzi Huohua Fenxi Dui Shang Shiqi Yuanshici Chandi De Yanjiu (a Study on the Provenance of Shang Protoporcelain by the Instrumental Neutron Activition Analysis). *Kaogu* 7.

Chen, Tiemei, George Rapp and Zhichun Jing, 1999, Provenance Studies of the Earliest Chinese Protoporcelain Using Instrumental Neutron Activation Analysis. *Journal of archaeological science* 26: 1003-1015.

2003, Shangzhou Shiqi Yuanshici De Zhongzi Huhua Fenxi Ji Xiangguan Wenti Taolun (a Discussion of the Instrumental Neutron Activition Analysis Analysis and Relevant Issues on Shang and Zhou Protoporcelain). *Kaogu* 7.

Chen, Xingcan, 1999, Zhongguo Shiqian Kaoguxue Yanjiu De Xinlu Licheng (on the History of the Study of Prehistoric Archaeology in China). In *Liangzhu Wenhua Yanjiu*, edited by Zhejiang Institute of Archaeology. Beijing, Kexue Press: 141-149.

Clark, John E., 1995, Craft Specialization as an Archaeological Category. *Research in economic anthropology* 16: 267-294.

Clark, John E. and W. Parry, 1990, Craft Specialization and Cultural Complexity. *Research in economic anthropology* 12: 289-346.

Costin, Cathy L., 1991, Craft Specialization: Issues in Defining, Documenting, and Explaining the Organization of Production. In *Archaeological Method and Theory (3)*, edited by Michael B. Schiffer. Tucson, The University of Arizona Press. 3: 1-56.

1996, Craft Production and Mobilization Strategies in the Inka Empire. In *Craft Specialization and Social Evolution: In Memoty of V. Gordon Childe*, edited by Bernard Wailes. Philadelphia, University of Pennsylvania: 211-222.

1998a, Housewives, Chosen Women, Skilled Men: Cloth Production and Social Identity in the Late Prehispanic Andes. In *Craft and Social Identity*, edited by Cathy Lynne Costin and Rita P. Wright. Arlington, Virginia, the American Anthropological Association. No. 8: 123-141.

1998b, Introduction: Craft and Social Identity. In *Craft and Social Identity*, edited by Cathy L. Costin and Rita P. Wright. Washington D.C., Archeological Papers No. 8. American Anthropological Association: 3-18.

2001, Craft Production Systems. In *Archaeology at the Millennium: A Sourcebook*, edited by Gary M. Feinman and T. Douglas Price. New York, Kluwer Academic/Plenum Publishers: 273-328.

Costin, Cathy L. and Melissa B. Hagstrum, 1995, Standardization, Labor Investment, Skill, and the Organization of Ceramic Production in Late Prehispanic Highland Peru. *American antiquity* 60 (4): 619-639.

Cross, J., 1993, Craft Specialization in Nonstratified Societies. *Research in economic anthropology* 14: 61-84.

Crumley, Carole L., 1995, Heterarchy and the Analysis of Complex Societies. In *Heterarchy and the Analysis of Complex Societies*, edited by R.M. Ehrenreich et al., Archaeological Papers of the American Anthropological Association Number 6: 1-6.

Dai, Xiangming, 1996, Peiligang Mudi Xintan (the New Research on the Peiligang Cemetery). *Huaxia kaogu* 3: 65-79.

1998a, Huanghe Liuyu Xinshiqi Shidai Wenhua Geju Zhi Yanbian (Evolution of the Neolithic Cultural Pattern in the Yellow River Valley). *Kaogu xuebao* 4: 389-418.

1998b, Shilun Miaodigou Wenhua De Qiyuan (a Discussion on the Origin of Miaodigou Culture). In *Qingguoji*, edited by Department of Archaeology of Jilin University. Beijing, Zhishi Press: 18-26.

2001, Zhongguo Kaoguxue Guannian De Huigu Yu Fansi (Review and Introspection on the Idea of Chinese Archaeology). *Dongnan wenhua* 1: 18-24.

2002, Miaodigou Wenhua De Shikong Jiegou (the Spatiotemporal Framework of Miaodigou Culture), Unpublished paper.

2004, Miaodigou Wenhua De Juluo Yu Shehui (the Settlements and Societies in the Miaodigou Culture). In *Gudai Wenming (Ancient Civilizations) (3)*, edited by Center for the Study of Chinese Archaeology & Aurora Centre for the Study of Ancient Civilizations in Beijing University. Beijing, Wenwu Press: 15-39.

Davis, Whitney M., 1981, Comment on P. Rice, Evolution of Specialized Pottery Production: A Trial Model. *Current Anthropology* 22: 228-230.

Deng, Zequn and Jiazhi Li, 1996, Jinnan Yuanqu Shangcheng Yizhi Gutaoci Huaxue Zucheng Ji Gongyi De Yanjiu (a Study About Chemical Composition and Manufacturing Technology of the Ancient Ceramics at the Yuanqu Shang Walled Town). In *Yuanqu Shangcheng*, edited by National Museum of Chinese History, Shanxi Institute of Archaeologyet al. Beijing, Kexue Press: 301-308.

Du, Jinpeng, Xuerong Wang and Liangren Zhang, 1999, Shilun Yanshi Shangcheng Xiaocheng De Jige

Wenti (on Several Issues Related to the Small City inside the Yanshi Shang City). *Kaogu* 2: 35-40.

Earle, Timothy K., 1977, A Reappraisal of Redistribution: Complex Hawaiian Chiefdoms. In *Exchange Systems in Prehistory*, edited by Timothy K. Earle and Jonathan E. Ericson. New Yord, Academic Press: 211-229.

1978, *Economic and Social Organization of a Complex Chiefdom: The Halelea District, Kaua'i, Hawaii.* Ann Arbor.

1984, Introduction. In *On the Evolution of Complexity: Essays in Honor of Harry Hoijer, 1982*, edited by Timothy K. Earle. Malibu, Undena Publications.

1987, Specialization and the Production of Wealth: Hawaiian Chiefdoms and the Inka Empire. In *Specialization, Exchange, and Complex Societies*, edited by E. M. Brumfiel and T. K. Earle. Cambridge, Cambridge University Press: 64-75.

1991, The Evolution of Chiefdom. In *Chiefdoms: Power, Economy, and Ideology*, edited by Timothy Earle. Cambridge, Cambridge University Press: 1-15.

1997, *How Chiefs Come to Power: The Political Economy in Prehistory.* Stanford, Stanford University Press.

Falconer, Steven E. and Stephen H. Savage, 1995, Heartlands and Hinterlands: Alternative Trajectories of Early Urbanization in Mesopotamia and the Southern Levant. *American Antiquity* 60(1): 37-58.

Fan, Wenquan, 2000, Guanjia Yizhi Fajue Huo Zhongyao Chengguo (the Important Achievement of Excavations at the Guanjia Site). *Zhongguo wenwubao, Feb. 13, 2000.* Beijing.

Feinman, Gary, 1995, The Emergence of Inequality: A Focus on Strategies and Processes. In *Foundations of Social Inequality*, edited by Douglas Price and Gary Feinman. New York, Plenum Press: 225-279.

Feinman, Gary, Stephen Kowalewski and Richard Blanton, 1984, Modelling Ceramic Production and Organizational Change in the Pre-Hispanic Valley of Oaxaca, Mexico. In *The Many Dimensions of Pottery: Ceramics in Archaeology and Anthropology*, edited by S.E. van der Leeuw and A. Pritchard. Amsterdam, Albert Egges van Giffen. Cingvla: 295-338.

Feinman, Gary and Jill Neitzel, 1984, Too Many Types: An Overview of Prestate Societies in the Americas. In *Advances in Archaeological Method and Theory*, edited by Michael B. Schiffer. New York, Academic Press. 7: 39-102.

Flad, Rowan Kimon, 2004, *Specialized Salt Production and Changing Social Structure at the Prehistoric Site of Zhongba in the Eastern Sichuan Basin, China.* Ph.D dissertation, University of California, Los Angeles.

Flannery, Kent V., 1972, The Cultural Evolution of Civilizations. *Annual Review of Ecology and Systematics* 3: 399-426.

Frankel, David, 1988, Pottery Production in Prehistoric Bronze Age Cyprus: Assessing the Problem. *Journal of Mediterranean archaeology* 1/2: 27-55.

Fried, Morton, 1967, *The Evolution of Political Society.* New York, Random House.

Gansu Committee of Cultural Relics Management, 1957, Lanzhou Xinshiqi Shidai De Wenhua Yicun (the Cultural Remains of the Neolithic at Lanzhou). *Kaogu Xuebao* 1: 1-8.

Gansu Cultural Relics Working Team, 1986, Gansu Qin'an Dadiwan 901 Hao Fangzhi Fajue Jianbao (Brief Report on the Excavation of the No.901 Structural Remains at Dadiwan in Qin'an, Gansu). *Wenwu* 2: 1-12.

Gansu Provincial Museum, 1983, Gansu Qin'an Dadiwan Yizhi 1978 Zhi 1982 Nian Fajue De Zhuyao Shouhuo (the Major Gains of Excavation at Dadiwan Site in Qin'an, Gansu, from 1978 to 1982). *Wenwu* 11: 21-29.

Gao, Wei, Tianlin Gao and Daihai Zhang, 1983, Guanyu Taosi Mudi De Jige Wenti (Some Issues About the Taosi Burial Site). *Kaogu* 6: 531-536.

Gao, Wei, Xizhang Yang, Wei Wang, Jinpeng Du, 1998, Yanshi Shangcheng Yu Xiashang Wenhua Fenjie. *Kaogu* 10: 66-79.

Hagstrum, Melissa B., 1985, Measuring Prehistoric Ceramic Craft Specialization: A Test Case in the American Southwest. *Journal of Field Archaeology* 12: 65-76.

Hassan, Fekri, 1981, *Demographic Archaeology.* New York, Academic Press.

Hayden, Brian, 2001, Richman, Poorman, Beggerman, Chief: The Dynamics of Social Inequality. In *Archaeology at the Millennium: A Sourcebook*, edited by Gary M. Feinman and T. Douglas Price. New York, Kluwer Academic/Plenum Publishers: 231-272.

He, Nu and Zhibin Yan, 2002, Huanghe Liuyu Shiqian Zuida Chengzhi Jinyibu Tanming (New Investigation of the Largest Prehistoric Walled Site in the Yellow River Valley). *Zhongguo Wenwubao.* Feb. 8.

He, Nu, Zhibin Yan and Jianzhong Song, 2003, Xiangfen Taosi Chengzhi Fajue Chengxian Baoli Secai (the Evidence of Violence in Taosi , Xiangfen). *Zhongguo Wenwubao.* Jan. 31.

Henan 1st Team, Institute of Archaeology CASS and Henan Institute of Cultural Relics, 2001, Henan Lingbaoshi Xipo Yizhi Shijue Jianbao (Brief Report of Excavation at the Xipo Site in Lingbao, Henan). *Kaogu* 11: 3-14.

Henan Institute of Cultural Relics, 1994, *Henan Kaogu Sishi Nian (Fourty Years of Archaeology in Henan).* Zhengzhou, Henan Renmin Press.

1999, *Wuyang Jiahu.* Beijing, Kexue Press.

2001, *Zhengzhou Shangcheng.* Beijing, Wenwu Press.

Henan Institute of Cultural Relics, Institute of Archaeology CASS, Sanmenxia City Cultural Relics Team and Lingbao Archaeological Team, 1999, Henan Lingbao Zhudingyuan Jiqi Zhouwei Kaogu Diaocha Baogao (Report of Surveys at Zhudingyuan and Surrounding Areas in Lingbao, Henan). *Huaxia kaogu* 3: 19-42.

2002, Henan Lingbaoshi Xipo Yizhi 2001 Nain Chun Fajue Jianbao (Brief Report of the Excavation in 2001 Spring at Xipo in Lingbao, Henan). *Huaxia Kaogu* 2: 31-52, 92.

2003, Henan Lingbao Xipo Yizhi 105 Hao Yangshao Wenhua Fangzhi(No. 105 House Foundation Dating to the Yangshao Culture at Xipo Site in Lingbao, Henan). *Wenwu* 8: 4-17.

Hill, J. N., W. N. Trierweiler, and R. W. Preucel, 1996, The evolution of cultural complexity: A case from the Pajarito Plateau, New Mexico. In *Emergent Complexity: The Evolution of Intermediate Societies*, edited by J. E. Arnold, pp. 107-127. International Monographs in Prehistory, Ann Arbor, Michigan.

Hodder, Ian, 1979, *The Spatial Organization of Culture.* Pittsburgh, University of Pittsburgh Press.

1981, Comment on P. Rice, Evolution of Specialized Pottery Production: A Trial Model. *Current Anthropology* 22: 231-232.

Inner Mongolia Institute of Cultural Relics, 2000, *Daihai Kaogu (1) (Archaeological Excavations at Daihai [I])*. Beijing, Kexue Press.

Inner Mongolia Team, Institute of Archaeology CASS, 1997, Aohanqi Xinglongwa Juluo Yizhi 1992 Nian Fajue Jianbao (Preliminary Report of the 1992 Season of Excavations at the Xinglongwa Site , Aohan Banner, Inner Mongolia). *Kaogu* 1: 1-26.

Institute of Archaeology, CASS, 1983, *Baoji Beishouling.* Beijing, Wenwu Press.

1999a, Survey at the Beiyangping Site in Lingbao City. *Kaogu* 12: 1-15.

1999b, *Yanshi Erlitou (Erlitou in Yanshi).* Beijing, Zhongguo Dabaikequanshu Press.

2001, *Mengcheng Yuchisi (Yuchisi in Mengcheng).* Beijing, Kexue Press.

Institute of Archaeology, Chinese Academia Sinica, 1963, *Xi'an Banpo.* Beijing, Wenwu Press.

Johnson, Allen W. and Timothy Earle, 2000, *The Evolution of Human Societies: From Foraging Group to Agrarian State.* Stanford, Stanford University Press.

Johnson, Gregory A., 1975, Locational Analysis and the Investigation of Uruk Local Excahnge Systems. In *Ancient Civilization and Trade*, edited by Jeremy A. Sabloff and C.C. Lamberg-Karlovsky. Albuquerque, University of New Mexico Press: 285-339.

1977, Aspect of Regional Analysis in Archaeology. *Annual Review of Anthropology* 6: 479-508.

1980, Rank-Size Convexity and System Integration: A View from Archaeology. *Economic Geography* 56(3): 234-47.

1981, Monitoring Complex System Integration and Boundary Phenomena with Settlement Size Data. In *Archaeological Approaches to the Study of Complexity,*, edited by Sander E. van der Leeuw. Amsterdam, A. E. van Giffen Instituut woor Prae- en Protohistorie: 143-188.

1987, The Changing Organization of Uruk Administration on the Susiana Plain. In *The Archaeology of Western Iran: Settlement and Society from Prehistory to the Islamic Conquest,*, edited by Frank Hole. Washington D.C., Smithsonian Institution Press: 107-140.

Keightley, David, 1983, The Late Shang State: When, Where, and What? In *The Origins of Chinese Civilization*, edited by David Keightley. Berkeley, University of California Press: 523-564.

Lass, Barbara, 1998, Crafts, Chiefs, and Commoners: Production and Control in Precontact Hawai'i. In *Craft and Social Identity*, edited by Cathy Lynne Costin and Rita P. Wright. Arlington, Virginia, the American Anthropological Association. No. 8: 19-30.

Lewis, Brandon S., 1996, The Role of Attached and Independent Specialization in the Development of Sociopolitical Complexity. *Research in economic anthropology* 17: 357-388.

Li, Fei , Shuicheng Li and Tao Shui, 1993, Hulu He Liuyu De Guwenhua Yu Guhuanjing (the Ancient Cultures and Environment in the Hulu River Valley). *Kaogu* 9: 822-842.

Li, Wenjie, 1996, Yuanqu Shangcheng Taoqi De Zhizuo Gongyi (the Manufacturing Technology of Pottery of the Yuanqu Shang Walled Town). In *Yuanqu Shangcheng*, edited by National Museum of Chinese History, Shanxi Institute of Archaeologyet al. Beijing, Kexue Press: 309-315.

2001, Yuanqu Gucheng Dongguan Yizhi Zhitao Gongyi Yanjiu (a Study on the Pottery Manufacture in Gucheng Dongguan, Yuanqu). In *Yuanqu Gucheng Dongguan*, edited by National Museum of Chinese History, Shanxi Institute of Archaeologyet al. Beijing, Kexue Press: 532-569.

Li, Xueqin (editor), 1997, *Zhongguo Gudai Wenming yu Guojia Xingcheng Yanjiu.* Kunming, Yunnan Renmin Press.

Liu, Li, 1994, *Development of Chiefdom Societies in the Middle and Lower Yellow River Valley in Neolithic China -- a Study of the Longshan Culture from the Perspective of Settlement Patterns.* Ph.D dissertation, Harvard University, Cambridge, Mass.

1996, Settlement Patterns, Chiefdom Variability, and the Development of Early States in North China.

Journal of anthropological archaeology 15: 237-288.
2000, The Development and Decline of Social Complexity in China: Some Environmental and Social Factors. *Indo-Pacific Prehistory Association Bulletin (Maelaka Papers) 20* 4: 14-33.
2003, "the Products of Minds as Well as of Hands": Production of Prestige Goods in the Neolithic and Early State Periods of China. *Asian Perspectives* 42(1): 1-40.
2004, *The Chinese Neolithic: Trajectories to Early States*. Cambridge, Cambridge University Press.
Liu, Li and Xingcan Chen, 2000, Cheng: Xia Shang Shiqi Dui Ziran Ziyuan De Kongzhi Wenti (Cities: Control of Natural Resources in the Xia and Shang Period). *Dongnan wenhua* 3: 45-60.
2001a, China. In *Encyclopedia of Archaeology: History and Discoveries*, edited by Tim Murray. Santa Barbara, ABC-CLIO. I, A-D: 315-333.
2001b, Settlement Archaeology and the Study of Social Complexity in China. *The Review of Archaeology* 22(2): 4-21.
2002, Zhongguo Zaoqi Guojia De Xingcheng: Erlitou He Erligang Shiqi Zhongxin He Bianyuan Diqu Zhijian De Guanxi (State Formation in Early China: Core-Periphery Relationships in the Erlitou and Erligang Periods). In *Gudai Wenming (Ancient Civilizations) (1)*, edited by Center for the Study of Chinese Archaeology & Aurora Centre for the Study of Ancient Civilizations in Beijing University. Beijing, Wenwu Press. 1: 71-136.
2003, *State Formation in Early Chna*. London, Duckworth.
Longacre, William A., 1999, Standardization and Specialization: What's the Link. In *Pottery and People: A Dynamic Interaction*, edited by J. Skibo and G. Feinman. Salt Lake city, The University of Utah Press: 44-58.
Longacre, William A., K. L. Kvamme and M. Kobayashi, 1988, Southwestern Pottery Standardization: An Ethnoarchaeological View from Philippines. *Kiva* 53: 101-112.
Ma, Xiaoling, 2003, *Emergent Social Complexity in the Yangshao Culture: Analyses of Settlement Patterns and Faunal Remains from Lingbao, Western Henan, China*. Ph.D dissertation, La trobe University, Melbourne.
Mou, Yongkang and Xizheng Yun, Eds, 1992, Zhongguo Yuqi Quanji (Selected Work of Chinese Jades). Shijiazhuang, Hebei Meishu Press.
National Museum of Chinese History, Shanxi Institute of Archaeology and Yuanqu County Museum, 1996, *Yuanqu Shangcheng: 1985-1986 Niandu Kancha Baogao (Yuanqu Shang City: Report of Excavations in 1985-1986)*. Beijing, Kexue Press.
1997a, 1988-1989 Nian Shanxi Yuanqu Gucheng Nanguan Shangdai Chengzhi Fajue Jianbao (Excavations of the Shang City Site at Nanguan, Gucheng, Yuanqu, Shanxi in 1988-1989). *Wenwu* 10: 12-29.
1997b, 1991-1992 Nian Shanxi Yuanqu Gucheng Nanguan Shangdai Chengzhi Fajue Jianbao (Excavations of the Shang City Site at Nanguan, Gucheng, Yuanqu, Shanxi in 1991-1992). *Wenwu* 12: 4-15.
2001, *Yuanqu Gucheng Dongguan (Gucheng Dongguan in Yuanqu)*. Beijing, Kexue Press.
Nordbeck, S., 1971, Urban Allometric Growth. *Geografiska Annaler* 53B: 54-67.
Peacock, D.P.S., 1982, *Pottery in the Roman World: An Ethnoarchaeological Apporach*. London, Longman.
Peebles, Christopher and Susan Kus, 1977, Some Archaeological Correlates of Ranked Societies. *American antiquity* 42(3): 421-448.
Pool, Christopher A. and Robert S. Santley, 1992, Middle Classic Pottery Economics in the Tuxtla Mountains, Southern Veracruz, Mexico. In *Ceramic Production and Distribution: An Integrated Approach*, edited by George J. Bey III and Christopher A. Pool. Boulder, Westview Press: 205-234.
Railey, Jim, 1999, *Neolithic to Early Bronze Age Sociopolitical Evolution in the Yuanqu Basin, North-Central China*. Ph.D dissertation, Washington University, Saint Louis.
Renfrew, Colin, 1974, Beyond a Subsistence Economy: The Evolution of Social Organization in Prehistoric Europe. In *Reconstructing Complex Societies, Supplement to the Bulletin of the American Schools of Oriental Research*, edited by Charlotte B. Moore. 20: 69-95.
Renfrew, Colin and Paul Bahn, 2000, *Archaeology: Theories, Methods and Practice*. London, Thames and Hudson.
Rice, Prudence M., 1981, Evolution of Specialized Pottery Production: A Trial Model. *Current Anthropology* 22(33): 219-240.
1984, The Archaeological Study of Specialized Pottery Production: Some Aspects of Method and Theory. In *Pots and Potters: Current Approaches in Ceramic Archaeology*, edited by Prudence M. Rice. Los Angeles, Monograph XXIV. Institute of Archaeology, University of California: 45-54.
1987, *Pottery Analysis: A Sourcebook*. Chicago, The University of Chicago Press.
1989, Ceramic Diversity, Production, and Use. In *Quantifying Diversity in Archaeology*, edited by R.D. Leonard and G.T. Jones. Cambridge, Cambridge University Press: 109-117.
1991, Specialization, Standardization, and Diversity: A Retrospective. In *The Ceramic Legacy of Anna O. Shepard*, edited by Ronald L. Bishop and Frederick W. Lange. Niwot, University Press of Colorado: 257-279.

1996, Recent Ceramic Analysis: 2. Composition, Production and Theory. *Journal of archaeological research* 4(3): 165-202.
Roux, Valentine, 2003, Ceramic Standardization and Intensity of Production: Quantifying Degrees of Specialization. *American antiquity* 68(4): 768-782.
Sabloff, Jeremy A. and Wendy Ashmore, 2001, An Aspect of Archaeology's Recent Past and Its Relevance. In *Archaeology at the Millennium: A Sourcebood*, edited by Gary M. Feinman and T. Douglas Price. New York, Kluwer Academic / Plenum Publishers: 11-37.
Sanders, William T., 1984, Pre-Industrial Demography and Social Evolution. In *On the Evolution of Complex Societies: Essays in Honor of Harry Hoijer*, edited by Timothy Earle. Malibu, Undena Publication: 7-39.
Santley, Robert S., 1994, Specialised Commodity Production in and around Matacapan: Testing the Goodness of Fit of the Regal-Ritual and Administrative Models. In *Archaeological Views from the Countryside: Village Communities in Early Complex Societies*, edited by Glean M. Shwartz and Steven E.Falconer. Washington, Smithsonian Institution Press: 91-108.
Service, Elman, R., 1962, *Primitive Social Organization.* New York, Random House.
Shanxi Institute of Archaeology, 1998, Yuanqu Ningjiapo Taoyaozhi Fajue Jianbao (Brief Report on Pottery Kilns Excavated at Ningjiapo, Yuanqu, Shanxi). *Wenwu* 10: 28-32.
Shanxi Institute of Archaeology and Ruicheng County Museum, 2002, Shanxi Ruicheng Qingliangsi Mudi Yuqi (Jade Items Discovered at the Qingliangsi Cemetery, Ruicheng, Shanxi). *Kaogu yu wenwu* 5: 3-6.
Shanxi Team, Institute of Archaeology CASS, 1983, 1978-1980 Nian Shanxi Xiangfen Taosi Mudi Fajue Jianbao (Report of the Excavation at the Taosi Cemetery in Xiangfen, Shanxi, in 1978-1980). *Kaogu* 1: 30-42.
1985, Shanxi Yuanqu Guwenhua Yizhi De Diaocha (Surveys of Ancient Sites in Yuanqu, Shanxi). *Kaogu* 10: 875-884.
1986a, Shanxi Yuanqu Longwang'ai Yizhi De Liangci Fajue (Two Excavations at the Longwang'ai Site, Yuanqu, Shanxi). *Kaogu* 2: 97-111.
1986b, Taosi Yizhi 1983-1984 Nian Iii Qu Juzhuzhi Fajue De Zhuyao Shouhuo (Major Gains from the Excavation of the Residential Area in Section Iii at the Taosi Site in 1983-1984). *Kaogu* 9: 773-781.
1987, Shanxi Yuanqu Fengcun Xinshiqi Shidai Yizhi De Fajue (Excavation of the Neolithic Site at Fengcun, Yuanqu, Shanxi). *Kaoguxue jikan* 5: 27-60.
1998, Shanxi Yuanquxian Xiaozhao Xinshiqi Shidai Yizhi De Shijue (a Trial Excavation of the Neolithic Site at Xiaozhao in Yuanqu, Shanxi). *Kaogu* 4: 7-15.
2001, Shanxi Yuanqu Xiaozhao Yizhi 1996 Nian Fajue Baogao (Report of the Excavation at the Xiaozhao Site in Yuanqu, Shanxi, in 1996). *Kaogu xuebao* 2: 189-226.
Shanxi Team, Institute of Archaeology CASS and Shanxi Linfen Cultural Bureau, 2003, Shanxi Xiangfenxian Taosi Yizhi I I Qu Juzhuzhi 1999-2000 Nian Fajue Jainbao. *Kaogu* 3: 3-17.
Shelach, Gideon, 1997, A Settlement Pattern Study in Northeast China: Results and Potential Contributions of Western Theory and Methods to Chinese Archaeology. *Antiquity* 71: 114-127.
1999, *Leadership Strategies, Economic Activity, and Interregional Interaction : Social Complexity in Northeast China.* New York, Kluwer Academic/Plenum.
Sima, Qian, 1982 [org. c. 100 B.C.], *Shiji (Historical Memoirs).* Beijing, Zhonghuashuju Press.
Sinopoli, Carla M., 1988, The Organization of Craft Production at Vijayanagara, South India. *American anthropologist* 90: 580-597.
1991, *Approaches to Archaeological Ceramics.* New York, Plenum Press.
1998, Identity and Social Action among South India Craft Producers of the Vijayanagara Period. In *Craft and Social Identity*, edited by Cathy Lynne Costin and Rita P. Wright. Arlington, Virginia, the American Anthropological Association. No. 8: 161-172.
2002, Learning About the Past through Archaeological Ceramics: An Example from Vijayanagara, India. In *Archaeology: Original Readings in Method and Practice*, edited by Peter N. Peregrine, Carol R. Emberet al. New Jersey, Pearson Education, Inc: 62-80.
Spielmann, Katherine, 1998, Ritual Craft Specialists in Middle Range Societies. In *Craft and Social Identity, Archeological Papers of the American Anthropological Association*, edited by Cathy L. Costin and Rita P. Wright. Arlington, American Anthropological Association. 8: 153-160.
Stark, Barbara L., 1985, Archaeological Identification of Pottery Production Locations: Ethnoarchaeological and Archaeological Data in Mesoamerica. In *Decoding Prohistoric Ceramics*, edited by B. Nelson. Carbondale, Southern Illinois University Press: 158-194.
1992, Ceramic Production in Prehistoric La Mixtequilla, South-Central Veracruz, Mexico. In *Ceramic Production and Distribution: An Integrated Approach*, edited by George J. Bey III and Christopher A. Pool. Boulder, Westview Press: 175-204.
1995a, Problems in the Analysis of Standardization and Specialization in Pottery. In *Ceramic Production in the American Southwest*, edited by B. Mills and P. Crown. Tucson, University of Arizona Press: 231-267.

Stark, Miriam t., 1995b, Economic Intensification and Ceramic Specialization in the Philippines: A View from Kalinga. *Research in Economic Anthropology* 16: 179-226.

Stein, Gil, 1996, Producers, Patrons, and Prestige: Craft Specialists and Emergent Elites in Mesopotamia from 5500-3100 B.C. In *Craft Specialization and Social Evolution: In Memory of V. Gordon Childe*, edited by Bernard Wailes. Philadelphis, University of Pennsylvania: 26-38.

Stein, Gil J. and M. James Blackman, 1993, The Organizational Context of Specialized Craft Production in Early Mesopotamian States. *Research in Economic Anthropology* 14: 29-59.

Steponaitis, Vincas, 1978, Location Theory and Complex Chiefdoms. In *Mississippian Settlement Systems*, edited by Bruce Smith. New York, Academic: 417-453.

Steward, Julain H., 1955, *Theory of Culture Change: The Methdology of Multilinear Evolution*. Urbana, University of Illinois Press.

Steward, Julain H. and Louis C. Faron, 1959, *Native People of South America*. New York, McGraw-Hill.

Su, Bingqi and Weizhang Yin, 1981, Guanyu Kaoguxue Wenhua De Quxi Leixing Wenti (on the Issue of the Distribution and Development of Regional Cultures in Chinese Archaeology). *Wenwu* 5: 10-17.

Thorp, Robert, 1991, Erlitou and the Search for the Xia. *Early China* 16: 1-38.

Tong, Weihua, 1998, Shangdai Qianqi Yuanqu Pendi De Tongzhi Zhongxin -- Yuanqu Shangcheng (the Early-Shang Political Center in the Yuanqu Basin -- Yuanqu Shang Town). *Zhongguo lishi bowuguan guankan* 1: 89-100.

Tosi, Maurizio, 1984, The Notion of Craft Specialization and Its Representation in the Archaeological Record of Early States in the Turanian Basin. In *Marxist Perspectives in Archaeology*, edited by Matthew Spriggs. New York, Cambridge University Press: 22-52.

Trigger, Bruce G., 1968, The Determinants of Settlement Patterns. In *Settlement Archaeology*, edited by Kwang-chih Chang. California, National Press Books: 53-78.

Underhill, Anne, 1990, *Changing Patterns of Pottery Production During the Longshan Period of Northern China, Ca. 2500-2000 B.C.* Ph.D dissertation, University of British Columbia, Vancouver, B.C., Canada.

1991, Pottery Production in Chiefdoms: The Longshan Period in Nothern China. *World Archaeology* 23(1): 12-27.

1994, Variation in Settlements During the Longshan Period of Northern China. *Asian Perspectives* 33(2): 197-228.

1996, Craft Production and Social Evolution During the Longshan Period of Northern China. In *Craft Specialization and Social Evolution: In Memmory of V. Gordon Childe*, edited by Bernard Wailes. Philadelphia, The University of Museum of Archaeology and Anthropology, University of Pennsylvania: 133-150.

2002, *Craft Production and Social Change in Northern China*. New York, Kluwer Academic/Plenum Publishers.

Underhill, Anne , G. Feinman, G. Nicholas, G. Bennet, F. Cai, H. Yu, F. Luan and F. Hui, 1998, Systematic, Regional Survey in Se Shandong Province, China. *Journal of Field Archaeology* 25: 453-474.

Underhill, Anne, Gary Feinman, Linda Nicholas, Gwen Bennett, Hui Fang, Fengshi Luan, Haiguang Yu and Fengshu Cai, 2002, Regional Survey and the Development of Complex Societies in Southeastern Shandong, China. *Antiquity* 76: 745-755.

Underhill, Anne P., 2003, Investigating Variation in Organization of Ceramic Production : An Ethnoarchaeological Study in Guizhou, China. *Journal of Archaeological Method and Theory* 10(3): 203-275.

van der Leeuw, Sander, 1977, Towards a Study of the Economics of Pottery Making. *Ex Horreo* 4: 68-76.

Wang, Lixin, 1998, *Zaoshang Wenhua Yanjiu*. Beijing, Gaodeng Jiaoyu Press.

Wattenmaker, Patricia, 1998, Craft Productioin and Social Identity in Northwest Mesopotamia. In *Craft and Social Identity*, edited by Cathy Lynne Costin and Rita P. Wright. Arlington, Virginia, the American Anthropological Association. No. 8: 47-56.

Wenke, R., 1991, The Evolution of Early Egyptian Civilization: Issues and Evidence. *Journal of World Prehistory* 5 (3): 279-329.

White, Joyee C., 1995, Incorporating Heterarchy into Theory on Socio-Political Development: The Case from Southeast Asia. In *Heterarchy and the Analysis of Complex Societies*, edited by R.M. Ehrenreich et al., Archaeological Papers of the American Anthropological Association, Number 6: 101-123.

White, Leslie A., 1959, *The Evolution of Culture: The Development of Civilization to the Fall of Rome*. New York, McGraw-Hill.

Willey, Gordon R., 1953, *Prehistoric Settlement Patterns in the Viru Valley, Bureau of American Ethnology Bulletin 155*. Washington, D.C., Smithsonian Institution.

Wright, Henry, 1977, Recent Research on the Origin of the State. *Annual review of anthropology* 6: 379-397.

1984, Prestate Political Formations. In *On the Evolution of Complex Societies: Essays in Honor of Harry Hoijer*, edited by T. Earle. Malibu, Undena Publication: 41-77.

1986, The Evolution of Civilizations. In *American Archaeology Past and Future*, edited by David J.

Meltzer, Don D. Fowleret al. Washington D. C., Smithsonian Institution Press.

Wright, Henry and Gregory Johnson, 1975, Population, Exchange, and Early State Formation in Southwestern Iran. *American Anthropologist* 77: 267-289.

Wright, Rita P., 1998, Crafting Social Identity in Ur Iii Southern Mesopotamia. In *Craft and Social Identity*, edited by Cathy Lynne Costin and Rita P. Wright. Arlington, Virginia, the American Anthropological Association. No. 8: 57-70.

Xi'an Banpo Museum, 1988, *Jiangzhai -- Xinshiqi Shidai Yizhi Fajue Baogao*. Beijing, Wenwu Press.

Xi'an Banpo Museum, Shaanxi Institute of Archaeology and Lintong County Museum, 1988, *Jiangzhai-- Xinshiqi Shidai Yizhi Fajue Baogao*. Beijing, Wenwu Press.

Xu, Xichen, Weizhi Si and Zhao Yang, Eds, 1994, Zhongguo Tongshi (3) (the Chinese History [Vol. 3]). *Zhongguo Tongshi (the Chinese History)*. Shouyi Bai. Shanghai, Shanghai Renmin Press.

Xu, Zhiyong, 1996, Guanyu Gucheng Dongguan Yangshao Zaoqi Wenhua Yicun De Ruogan Xiangguan Wenti (on Several Issues Relevant to the Cultural Remains of Early Yangshao at Dongguan, Gucheng). In *Zhongguo Lishi Bowuguan Kaogubu Jinian Wenji*, edited by Department of Archaeology of National Museum of Chinese History. Beijing, Kexue Press: 26-33.

Xue, Xinming, 2004, Shanxi Ruicheng Sili--Potou Yizhi Faxian Miaodigou Erqi Wenhua Mudi (a Discovery of the Miaodigou Ii Cemetery at Sili-- Potou, Ruicheng, Shanxi). *zhongguo wenwubao, April 16, 2004*. Beijing.

2005, Shanxi Ruicheng Qingliangsi Jielu Miaodigou Erqi Wenhua Daxing Mudi. *Zhongguo Wenwubao, Feb 23, 2005*. Beijing.

Yan, Wenming, 1981, Jiangzhai Zaoqi De Cunluo Buju (Settlement Layout of Jiangzhai in the Early Phase). *Kaogu yu wenwu* 1: 63-71.

1987, Zhongguo Shiqian Wenhua De Tongyixing Yu Duoyangxing (the Unity and Variability of Prehistoric Chinese Culture). *Wenwu* 3: 38-50.

1989a, Yangshao Fangwu He Juluo Xingtai Yanjiu (on the Architecture and Settlement Layout in the Yangshao Culture). In *Yangshao Wenhua Yanjiu*, edited by Wenming Yan. Beijing, Wenwu Press: 180-242.

1989b, Zhongguo Xinshiqi Shidai Juluo Xingtai De Kaocha (Investigation of Settlement Patterns in Neolithic China). In *Qingzhu Su Bingqi Kaogu Wushiwu Nian Lunwenji*, edited by "Qingzhu Su Bingqi kaogu wushiwu nian lunwenji" Editorial Group. Beijing, Wenwu Press: 24-37.

1992, Luelun Zhongguo Wenming De Qiyuan (Brief Discussion on the Origins of Chinese Civilization). *Wenwu* 1: 40-49.

1996, Zhongguo Wenming Qiyuan De Tansuo (an Exploration on the Origin of Chinese Civilization). *Zhongyuan wenwu* 1: 10-16.

1997, Huanghe Liuyu Wenming De Faxiang Yu Fazhan (the Origins and Development of Civilization in the Yellow River Valley). *Huaxia kaogu* 1: 49-54.

1999, Wenming Qiyuan Yanjiu De Huigu Yu Sikao (Review and Thinking on the Study of the Origins of [Chinese]Civilizations). *Wenwu* 10: 27-34.

2000, Zhongguo Huanhao Juluo De Yanbian (the Evolution of Chinese Fortified Settlements). In *Nongye Fasheng Yu Wenming Qiyuan (the Origins of Agriculture and the Rise of Civilisation)*, edited by Wenming Yan. Beijing, Kexue Press: 249-255.

Yang, Bojun, Ed. 1981, Chunqiu Zuozhuan Zhu. Beijing, Zhonghua Press.

Yang, Shengnan, 1992, *Shangdai Jingjishi*. Guiyang, Guizhou Renmin Press.

Yoffee, Norman, 1993, Too Many Chiefs? In *Archaeological Theory: Who Sets the Agenda?*, edited by Norman Yoffee and Andrew Sherratt. Cambridge, Cambridge University Press: 60-78.

Yuanqu County Chronicles Editorial Board, 1993, *Yuanqu Xianzhi*. Taiyuan, Shanxi Renmin Press.

Zhang, Chi, 2000, Daxi, Beiyinyangying He Xuejiagang De Shi, Yu Qi Gongye (Lithic and Jade Industries of the Daxi, Beiyinyangying and Xuejiagang Cultures). In *Kaoguxue Yanjiu (4)*, edited by Beijing University Department of Archaeology. Beijing, Kexue Press: 55-76.

2002, Zhongguo Shiqian Nongye Jingji De Fazhan Yu Wenming De Qiyuan (Development of Prehistoric Agriculture and Economy and the Origin of Civilizations in China). In *Gudai Wenming (1)*, edited by Center for the Study of Chinese Archaeology in Beijing University. Beijing, Wenwu Press. 1: 35-57.

2003, *Changjiang Zhongxiayou Diqu Shiqian Juluo Yanjiu (Prehistoric Settlement Pattern of the Mid and Lower Yangtze River Basin)*. Beijing, Wenwu Press.

Zhang, Sulin, 1995, Shilun Yuanqu Gucheng Dongguan Miaodigou Erqi Wenhua (a Discussion on Miaodigou Ii Culture at Dongguan in Gucheng, Yuanqu). *Wenwu jikan* 4: 38-48.

Zhao, Chunqing, 2001, *Zhengluo Diqu Xinshiqi Shidai Juluo De Yanbian (Change of Neolithic Settlements in the Zhengzhou-Luoyang Region)*. Beijing, Beijing University Press.

Zhao, Hui, 2000, Zhongguo Wenming Qiyuan Yanjiu Zhong De Yige Jiben Wenti (an Essential Issue in Studies of the Origions of Chinese Civilization). In *The Origions of Rice Agriculture, Pottery and Cities*, edited by Yan Wenming and Yasuda Yoshinori. Beijing, Wenwu Press: 135-142.

Zhu, Fenghan, 2004, *Shangzhou Jiazu Xingtai Yanjiu (a Study on the Pattern of Lineages of the Shang and Zhou Dynasties)*. Tianjin, Tianjin Guji Press.

Zhu, Yanping, 1988, Peiligang Wenhua Mudi Zaitan (the Re-Study on Cemeteries of Peiligang Culture). *Kaogu* 11: 1021-1034.

Zou, Heng, 1980, Lun Tangdou Zhengbo Jiqi Qianhou De Qianxi. In *Xiashangzhou Kaogu Lunwenji (a Collection of Papers on Xia, Shang and Zhou Archaeology)*, edited by Heng Zou. Beijing, Wenwu Press: 1183-218.

www.ingramcontent.com/pod-product-compliance
Lightning Source LLC
LaVergne TN
LVHW071219210926
848482LV00015B/848